POULTRY
MEAT AND EGG
PRODUCTION

POULTRY MEAT AND EGG PRODUCTION

Carmen R. Parkhurst

Department of Poultry Science
North Carolina State University
Raleigh, North Carolina

George J. Mountney

United States Department of Agriculture
Cooperative State Research Service
Washington, DC

An **avi** Book

CHAPMAN & HALL
New York • London

This edition published by
Chapman & Hall
One Penn Plaza
New York, NY 10119

Published in Great Britain by
Chapman & Hall
2-6 Boundary Row
London SE1 8HN

Printed in the United States of America

An AVI Book
Copyright © 1988 by Van Nostrand Reinhold

Please send your order for this or any **Chapman & Hall book to Chapman & Hall,
29 West 35th Street, New York, NY 10001, Attn: Customer Service Department.**
You may also call our Order Department at 1-212-244-3336 or fax your purchase order
to 1-800-248-4724.

For a complete listing of Chapman & Hall's titles, send your requests to **Chapman &
Hall, Dept. BC, One Penn Plaza, New York, NY 10119.**

16 15 14 13 12 11 10 9 8 7 6 5 4 3 2

Library of Congress Cataloging-in-Publication Data

Poultry meat and egg production.

   "An AVI book."
   Includes bibliography and index.
   1. Poultry.   2. Eggs—Production.   I. Parkhurst,
Carmen R. (Carmen Robert), 1942–
II. Mountney, George J.
SF487.P8755   1987   636.5   87-29590
ISBN 0-412-99741-X

To Dr. Robert E. Cook,
Former Head, Department of Poultry Science,
for his encouragement, patience, and understanding
during the time we wrote this book.

Contents

Preface   **xiii**

1   The Poultry Industry   **1**

    Introduction   1
    Biological Classification   1
    Domestication   2
    American Poultry Association Classification   3
    Commercial Classification   5
    Poultry as Food-Producing Animals   5
    Poultry as Laboratory Animals   6
    World Poultry Production   6
    The United States Poultry Industry   7
    Consumption of Poultry Products   11
    Careers in the Poultry Industry   13
    References   15

2   Anatomy and Structure of the Fowl   **16**

    Feathers   17
    Skin   18
    Muscles   19

Skeleton   19
Circulatory System   21
Body Temperature   22
Respiration   22
Digestion   23
Mouth and Esophagus   23
Crop and Proventriculus   24
Gizzard   24
Pancreas   25
Liver   25
Small Intestine   25
Ceca   27
Large Intestine, Rectum, and Cloaca   27
Excretory System   27
Nervous System   27
Avian Senses   28
Endocrine Organs   29
References   30

3 Physiology and Reproduction in Poultry   31

Endocrine System   31
Physiology of Reproduction   33
Reproduction   35
References   48

4 Genetics and Poultry Breeding   49

The Cell Theory   49
Cell Division   50
Gene Function   51
Genetic Code   52
Mutations   54
Phenotypic Expression (Nonadditive)   55
Phenotypic Expression of Genes (Additive)   56
Heritability Estimates   57
Selection   58
Current Breeding Systems   58
Strain Crosses   58
Crossbreeding   59
Selecting Superior Genetic Stock   59
Progeny Testing   60
Sources of Genetic Stock   60
Breeds and Varieties of Chickens   60
The Utilitarian Concept   61
Future Poultry Genetics and Breeding   62
Lethal Genes   63
Parthenogenesis   63
References   64

5 Incubation and Hatchery Management   65

Poultry Reproduction   65
Storage and Selection of Hatching Eggs   66
Embryonic Development   67

Extraembryonic Membranes   68
Daily Embryonic Growth   69
Embryonic Nutrition   73
Embryonic Communication   73
Embryology in Research and Teaching   74
Artificial Incubation   74
Modern Incubators   76
Temperature   76
Relative Humidity   77
Air Supply   78
Hatching Egg Positions   79
Embryonic Mortality   81
The National Poultry Improvement Plan   81
Hatchery Sanitation   82
Franchise Hatchery System   82
Hatchery Services   83
References   84

6 Social Behavior and Animal Welfare   85

Social Organization   85
Behavior of Feral Chickens   86
Communication   88
Behavior of Domestic Poultry   89
Social Behavior in Growing Chicks   90
Other Adult Social Behavior Characteristics   91
Animal Rights   92
The Humane Movement   93
Animal Welfarists   93
Animal Rightists   93
Animal Liberationists   94
Agricultural Involvement with Humane Groups   94
Public Relations   94
Need for a Code of Ethics   95
References   96

7 Environment and Housing   97

Farmstead Planning   99
External Services and Utilities   99
Water Supply   100
Poultry Building Orientation   100
Aesthetic Value   101
House Construction   102
Turnkey Houses   102
Criteria for Selecting a Manufactured House   102
Vapor Barriers   106
Ventilation   106
References   109

8 Poultry Nutrition   110

Nutrients   112
References   125

9   Diseases and Parasites of Poultry   126

Disease Prevention   127
Diagnostic Organs in Postmortem Examinations   128
Immunology of Disease Control   130
Poultry Disease Control Strategy   131
Vaccination   131
Nutritional Deficiency Diseases   132
Protozoan Diseases   133
Bacterial Diseases   137
Fowl Cholera   137
Nonrespiratory Bacterial Diseases   138
Mycoplasma   140
Viral Diseases   141
Fungal and Mold Diseases   144
Mycotoxicosis   145
External Parasites   146
Internal Parasites   148
Behavioral Diseases   149
Reproductive Diseases   150
References   150

10   Poultry and Egg Marketing   151

Assembling   151
Transporting   152
Assumption of Risk   152
Market News   153
Futures Trading   153
Processing Poultry   154
Further Processing   158
Eggs   159
Storage   160
Inspection and Grading   161
Packaging   168
Merchandising   170
New Product Development   172
References   172

11   Broiler Production   173

United States Broiler Industry   177
World Broiler Industry   177
Broiler Breeders   179
Housing and Egg Production   180
Care of Hatching Eggs   181
The Hatchery   181
Housing for Broilers   181
Litter   182
Brooding   182
Brooding Methods   183
Density   184
Placement   184
Management Suggestions for Partial House Brooding   185
Watering Equipment   186

Feeding Equipment   186
Lighting Programs   188
Ventilation   188
Gas Concentrations   188
Ventilation for Partial House Brooding   189
Fan Timers   189
Fan Thermostats   189
Air Intakes   190
Broiler Diets   190
Catching and Hauling Broilers   190
Grower Responsibilities   191
Catching Crews   191
Types of Broilers   191
Roasters   192
Grower Contracts   192
References   193

12   Commercial Egg Production   194

Starting Pullets in Cages   197
Brooding   198
Growing Pullets in Cages   199
Managing Laying Hens in Cages   201
Induced Molting   203
References   205

13   Turkey Production   206

Origin of the Turkey   206
Color Varieties   207
Turkey Genetics   208
Turkey Breeder Flocks   210
Turkey Breeders   210
Turkey Breeder Toms   211
Turkey Breeder Hens   211
Artificial Insemination   213
Hatching Eggs   213
Hatchery Services   214
Brooding   215
Floor Brooding   215
Floor Covering   216
Brooders   216
Brooder Guards   217
Equipment Arrangement   218
Trough Feeders   220
Mechanical Feeders   220
Waterers   220
Rearing Turkeys   220
References   226

14   Waterfowl Production   227

Ducks   227
Geese   236

Processing Waterfowl   245
Handling Waterfowl   247
Pinioning   247
Sexing Waterfowl   247
References   247

15 Miscellaneous Poultry   **249**

Coturnix Quail   249
Bobwhite Quail   253
Chukar Partridge   254
Pheasants   255
Guinea Fowl   257
Peafowl   259
Pigeons   261
Swan   263
References   265

16 Poultry Management Practices   **266**

Production Management   266
Specialized Procedures and Devices   271
Sanitation and Waste Disposal Procedures   276
Financial Management   284
References   285

Index   **287**

Preface

Poultry Meat and Egg Production has been prepared primarily for use as a text for students taking their first courses in poultry management. The general overall science and production practices currently in use in the industry have been characterized and described so that the student can gain insight into the industry. Reading portions of chapters before the lecture discussions and laboratory sessions will be helpful in giving students an understanding of the material. Also, this gives the instructor an opportunity to emphasize in the lectures areas of current concern in the industry, and to present topics of his or her choice in greater detail.

We wish to acknowledge and thank the following scientists who reviewed and critically evaluated the several chapters and made many helpful suggestions: Dr. Bobby Barnett, Clemson University; Mr. D. O. Bell, University of California; Dr. Donald Bray (retired), University of Illinois; Dr. W. H. Burke, University of Georgia; Dr. Frank Cherms, Nicholas Turkey Breeding Farms, Inc., Sonoma, California; Dr. Wendell Carlson (retired), South Dakota State University; Dr. J. V. Craig, Kansas State University; Dr. K. Goodwin (retired), Pennsylvania State University; Dr. T. L. Goodwin, University of Arkansas; Dr. G. C.

Harris, University of Arkansas; Dr. P. S. Hester, Purdue University; Dr. Glyde Marsh, Ohio State University; Dr. C. B. Ryan (retired), Texas A&M University; Dr. T. D. Siopes, North Carolina State University; Dr. L. D. Schwarz, formerly Pennsylvania State University, now Michigan State University; Dr. J. L. Skinner, University of Wisconsin; Dr. J. F. Stephens, Ohio State University; and Dr. J. P. Thaxton, North Carolina State University.

1

The Poultry Industry

INTRODUCTION

The term poultry is used collectively to designate those species of birds that have been domesticated to reproduce and grow in captivity and that render products of economic value. Chickens, turkeys, ducks, geese, some quail and pheasants, guineas, and pigeons generally meet the above criteria. They provide meat, eggs, feathers, fertilizer, animal food, and other by-products such as pharmaceuticals. They also serve as laboratory animals for scientific research. Birds kept only for companionship or beauty are not considered poultry.

BIOLOGICAL CLASSIFICATION

Chickens belong to the genus *Gallus* of the family Phasianidae. Chickens along with closely related families make up the galliformes or gallinaceous birds, which are terrestrial, chickenlike in appearance being heavy bodied, short duration flyers, scratching seed and insect feeders, and ground nesters with precocial (the young are hatched down covered and are able to feed and move about by themselves)

1

young. This group of birds (Galliformes) includes chickens, turkeys, guineas, peafowl, and quail to name the common ones.

The domestic chicken is *Gallus domesticus* and likely had its ancestry in the red jungle fowl *G. gallus* that originated in southeast Asia. Other species of wild jungle fowl include *G. sonnerati,* the gray jungle fowl, *G. lafayetti,* the Ceylonese jungle fowl, and *G. varius,* the green jungle fowl.

DOMESTICATION

It appears that people probably domesticated chickens over 4000 years ago, after centuries of hunting the wild jungle fowl for food. The early domesticated fowls were also used in religious ceremonies dedicated to the sun. In ancient India these chickens were sacrificed to the sun god.

Cocks were pitted together in fights originally as a kind of fertility ritual as an attempt by primitive peoples to ensure many children, bountiful crops, and adequate livestock.

Chickens then probably spread through eastern Asia. They reached Persia about 1000 B.C. and played a role in their ancient religion. By around 500 B.C. chickens were raised by the Greeks for the "sport" of cockfighting; however, the Romans were probably the first poultrymen. They used chickens in religious ceremonies, predicting the future through the reading of the livers by the horuspices. They also made prophecies by offering feed to sacred chickens. Hearty appetites predicted success; ignored food or listless pecking was a harbinger of failure and doom. A flock of sacred chickens became a part of many Roman general's field equipment. The chickens were fed or fasted

FIG. 1.1. Barred Plymouth Rocks. An example of the American class. (Drawing by J. L. Skinner.)

according to the whims of the general to control the morale of the troops.

The Romans also kept and bred chickens for food. They designed and built poultry houses and recognized the need for good sanitation. They were the first to fumigate poultry houses by using the fumes of burning pitch and sulfur.

AMERICAN POULTRY ASSOCIATION CLASSIFICATION

At the present time the American Poultry Association lists and classifies 300 different recognized breeds and varieties of chickens in its book, "The American Standard of Perfection." Most of these birds are kept for competition in poultry shows or as pets. They are classified into classes, breeds, and varieties.

Birds classified in the same class have a common origin. Those in the same class with the same general physical features such as body shape or type, skin color, number of toes, and feathered or unfeathered shanks are classified as breeds. Breeds are subdivided further into varieties, which are based on plumage color, comb type, and the presence or absence of a beard or muffs.

The American class (Fig. 1.1) includes many of the chickens commonly seen on farms before World War II. They were bred to forage and survive on farms with a minimum of care. Such birds as Rhode Island and New Hampshire Reds, White and Barred Plymouth Rocks, Wyandottes, and Jersey Giants are included in this class.

Most chickens in the Mediterranean class (Fig. 1.2) originated in Italy or Spain. Birds in this class are known for their active, nervous

FIG. 1.2. Blue Andalusians. An example of the Mediterranean class. (Drawing by J. L. Skinner.)

FIG. 1.3.   Dorkings. An example of the English class. (Drawing by J. L. Skinner.)

dispositions. The best known of these birds are the Leghorns, which originated in the city of Livorno (Leghorn), Italy. Less well known birds are Minorcas, Anaconas, Blue Andalusians, and White Faced Black Spanish.

Birds in the English class (Fig. 1.3) include the Orpingtons, Sussex, Austrolops, Dorkings, and Cornish. One of their chief characteristics, with the exception of Cornish which did not originate in England, is a very white skin.

Birds in the Asiatic class (Fig. 1.4) can be quickly identified since they are the only ones with feathers on their shanks. They have slow,

FIG. 1.4.   Brahmas (left) and Cochins (right). Examples of the Asiatic class. (Drawing by J. L. Skinner.)

lethargic temperaments. Examples of chickens in this class are Cochins, Brahmas, and Langshans.

The other classes did not play as important a part in the development of our modern chickens.

Bantams are miniature reproductions of larger birds. Typical examples are Barred, White Rock, and Rhode Island Red Bantams.

COMMERCIAL CLASSIFICATION

Chickens raised to produce eggs and meat and turkeys raised for meat make up by far the largest numbers of poultry raised and consumed.

Chickens are selected and bred to produce either large numbers of high quality eggs or vigorous rapid growing offspring with thick chunky breasts and meaty thighs. They are referred to as either egg-type or meat-type chickens. They are also classified as "white" and "brown" egg breeds or crosses. Turkeys selected for broad breasts and meaty thighs are produced for meat only.

POULTRY AS FOOD-PRODUCING ANIMALS

Poultry and especially chickens are now found in almost all parts of the world, regardless of climate. As nearly as can be estimated, there is now about one domestic chicken for every human being in the world and the chicken population explosion continues to gain on the human explosion.

Poultry can survive under very extreme temperatures. For example, guinea fowl survive and multiply in northern Siberia and chickens and ducks are frequently kept in the tropics.

Poultry enterprises, when compared with other livestock ventures, are easy and economical to establish. Within 7 weeks of hatching, 4 lb broilers can be processed for a meat supply and eggs can be produced in as little as 6 months.

Hatching eggs can be sent anywhere in the world without the need of sending a caretaker or feed or water with the eggs. Such eggs, when fertile, can be stored up to 2 weeks after they are laid before incubation is started. They can stand shipment without controlled temperatures as long as the eggs are kept from freezing and otherwise are maintained within the human comfort zone. In addition, by using eggs to populate a new poultry enterprise, the risk of introducing diseases that can occur with live birds is greatly reduced. Also, the newly hatched chicks either adapt and reproduce in the environment into which they were hatched or they die. Thus, adaptation begins immediately, as only survivors reproduce.

Several other advantages of poultry, especially for developing countries, include the fact that they can be "stored" on the farm as live birds and slaughtered a few at a time as needed. With larger animals, the slaughter of a hog or steer may require the immediate use of several hundred pounds of meat. Also, the low initial cost of eggs, production equipment, and housing enables low-income farmers to provide a source of meat protein with a low fixed investment.

Poultry have some ability to scavenge and use native feeds. They are also efficient converters of feed. Under ideal conditions with broilers, it requires slightly under 2 lb of feed for each pound of chicken that is produced. However, to attain this efficiency, broilers require a high protein ration and utilize some of the same food which could be used to feed humans.

POULTRY AS LABORATORY ANIMALS

Chickens and quail are widely used as laboratory animals, especially in feeding experiments. They are docile and grow well in cages where feed and water intake can be easily controlled and measured, they are easily handled, examined, and weighed, they can be raised in a small amount of space in large numbers, and they are quite sensitive to minor changes or deficiencies in diet.

Eggs and developing chick embryos are used for virus and vaccine culture. In the early 1940s it was found that biologicals for the pharmaceutical industry could be produced in embryonated chicken eggs instead of in living animals.

WORLD POULTRY PRODUCTION

The technological advances in poultry production developed in the United States and other countries in the last 40 years have been rapidly applied worldwide and have made it necessary to handle birds in large flocks. The need for high-quality nutrients, especially protein, and increasing levels of income and standards of living have created a tremendous demand for poultry products. Farm income from poultry alone in the United States amounted to $9.2 billion in 1982. Figure 1.5 indicates the main broiler-producing countries of the world and the percentages produced. The United States is the leading broiler producer in the world followed by Brazil, Japan, France, Spain, and the USSR. The United States produces over one-third (37%) of the world's broilers, has about one-sixth of the world's chickens, and contains only about 6% of the world's human population. For these reasons, whole

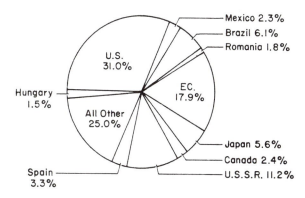

FIG. 1.5. Global poultry meat production—CY 1984. (Source: USDA 1985B.)

broilers sold for approximately $1/kg in the United States in 1983, $4.50 in Stockholm, $4.00 in Paris, $3.75 in Brussels, and $3.50 in Tokyo.

World trade in poultry meat is dominated by the European Economic Community, the United States, and Hungary. Chickens, broilers, and turkeys are the main types of meat produced. In the United States exports of poultry products, meat, eggs, and live poultry amounted to $409.1 million in 1979.

THE UNITED STATES POULTRY INDUSTRY

Because chickens can be hatched on a schedule, can be handled as a flock instead of individuals, can adapt to confinement, and because the cost of single birds is low, it is possible to use modern mass production methods to economically produce poultry meat and eggs on a continuous basis at prices within the reach of the average consumer. In fact, the industry is now so efficient that more food in the form of poultry products can be produced with less raw materials and energy consumption than is required from any other animal source other than some phases of aquaculture.

Poultry production has become specialized to the point that farmers contract with large integrated companies to produce broilers, eggs, and turkeys (Table 1.1). The companies supply the chicks, feed, litter, fuel, and medication and the contract grower provides housing and the labor for care of the birds. The contract growers receive a graduated fee which reflects the bird performance and management. Because capital requirements are so high for a poultry enterprise and the margin of profit per bird has become so small, individual growers cannot afford the financial risks associated with buying, growing, and selling poul-

TABLE 1.1. Changes in Integration of Production and Technical Efficiency Gains in Egg and Poultry Production in the United States for Selected Years[a]

	1955	1960	1965	1970	1975	1977
Market eggs						
Percentage of						
Contract production	0.5	7.0	18.0	20.0	37.0	44.0
Owner-integrated production	1.5	5.5	12.5	20.0	32.0	37.0
Contract marketing	12.5	13.5	13.5	15.0	10.0	8.0
Total	14.5	26.0	44.0	55.0	79.0	89.0
Pounds of feed/dozen	5.50	5.20	4.95	4.55	4.25	4.25
Commercial broilers						
Percentage of						
Contract production	87.0	90.0	90.0	90.0	90.0	88.0
Owner-integrated production	2.0	5.0	5.5	7.0	8.0	10.0
Contract marketing	1.0	1.0	1.5	2.0	1.0	1.0
Total	90.0	96.0	97.0	99.0	99.0	99.0
Pounds of feed/lb live weight	2.85	2.48	2.28	2.10	2.10	2.10
Market turkeys						
Percentage of						
Contract production	21.0	30.0	35.0	42.0	47.0	52.0
Owner-integrated production	4.0	4.0	8.0	12.0	20.0	28.0
Contract marketing	11.0	16.0	13.0	18.0	14.0	10.0
Total	36.0	50.0	56.0	72.0	81.0	90.0
Pounds of feed/lb live weight	4.40	3.90	3.50	3.25	3.10	3.10
Output per man-hour of labor in poultry production (1967 = 100)	32	55	87	120	175	215

[a] Source: Schertz (1979).

try on the open market. As a result of this system, the total poultry industry is linked or integrated either directly or indirectly to other industries such as transportation, financing, construction, and equipment manufacturers. Most large integrated operations have separate breeding farms, hatcheries, grow-out units, processing plants, and marketing and distribution facilities.

Since feed amounts to about 66–73% of the cost of producing a chicken (Table 1.2) the effect on United States grain production is quite profound. Over 45% of all commercially processed feed in the United States is fed to poultry. One of every five bushels of corn produced in the United States and 40% or more of the high protein feed fed to livestock is consumed by poultry. Figure 1.6 shows the functions of a typical integrated broiler firm.

Although poultry and eggs are produced all over the United States, most are produced in only a few concentrated areas. Broiler production

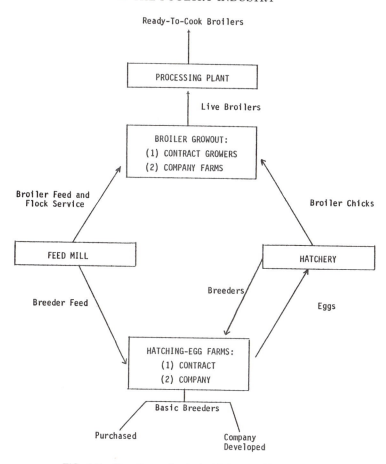

FIG. 1.6. Functions of a typical integrated broiler firm.

TABLE 1.2. Changes in Relative Importance of Production Input Costs in the United States for Selected Periods[a]

Item	Eggs (%)		Broilers (%)		Turkeys (%)	
	Mid-1960s	Mid-1970s	Mid-1960s	Mid-1970s	Mid-1960s	Mid-1970s
Feed	59	66	64	73	69	72
Hen depreciation	21	19	—	—	—	—
Chicks	—	—	18	12	—	—
Poults	—	—	—	—	15	11
Labor/mgt.	9	6	7	6.5	8	6
Energy	1	1	2	2	1	2
Other variables	3	2	4	2	3	2
Overhead	7	6	5	4.5	4	7
Total	100	100	100	100	100	100

[a] Source: Anonymous (1977A).

TABLE 1.3. Per Capita Consumption[a]

Year	Total eggs	Chickens and turkeys (lb)	All chickens (lb)	Broilers only (lb)	Turkeys (lb)	Retail weights (lb)			
						All red meats	Beef and veal	Pork	Lamb and mutton
1970	309.0	48.4	40.4	36.8	8.0	151.6	86.4	62.3	2.9
1971	310.6	48.6	40.3	36.5	8.3	156.7	85.6	68.3	2.8
1972	302.9	50.7	41.8	38.2	8.9	153.1	87.3	62.9	2.9
1973	289.2	49.0	40.5	37.2	8.5	141.7	82.0	57.3	2.4
1974	283.7	49.5	40.7	37.2	8.8	151.3	87.5	61.8	2.0
1975	276.4	48.6	40.1	36.7	8.5	143.8	91.3	50.7	1.8
1976	269.9	51.9	42.8	39.9	9.1	152.9	97.6	53.7	1.6
1977	267.6	53.3	44.2	41.1	9.1	152.2	94.9	55.8	1.5
1978	272.6	55.9	46.7	43.8	9.2	147.0	89.7	55.9	1.4
1979	277.7	60.5	50.6	47.7	9.9	144.8	79.7	63.8	1.3
1980	272.4	60.6	50.1	47.0	10.5	147.7	78.0	68.3	1.4
1981	265.6	62.4	51.6	48.6	10.8	145.2	78.8	65.0	1.4
1982	265.4	63.9	53.1	50.0	10.8	139.4	78.8	59.0	1.7
1983	262.0	65.1	53.8	50.8	11.3	144.0	80.4	62.2	1.5
1984[b]	261.0	67.1	55.7	53.0	11.4	143.5	80.3	61.6	1.5

[a] Source: USDA (1985A).
[b] Preliminary.

tends to be concentrated in the southeast. Turkey production is substantial in several areas around the country close to population centers, including North Carolina, California, and Minnesota. Egg production, while heavy in the southeast, has major concentrations close to large population centers such as California and Pennsylvania.

CONSUMPTION OF POULTRY PRODUCTS

Poultry meat is nutritious, tender, easy to chew or grind, mild in flavor, blends well with other foods, and is easy to handle and digest. It is available fresh and frozen year round and is quite economical when compared with other animal products. Primarily for the above reasons, consumption has been increasing steadily over a number of years (Table 1.3). Figure 1.7 shows the consumption trends for eggs and poultry meats.

Eggs are also nutritious, economical, easy to prepare, and easily digestible. After a steady decline in consumption since the 1940s, it appears the rate of decline in egg consumption has leveled off and consumption levels are stabilizing around 265 eggs per capita annually (Table 1.4).

It is predicted that consumption of poultry and eggs will continue to increase in the future with most of the increases in the form of new processed poultry products. In fact, the demand for processed poultry products has increased (Fig. 1.8) to the point where red meat packers are purchasing poultry processing operations. An example of the

TABLE 1.4. Per Capita Egg Consumption

Year	Shell	Processed	Total
1971	274.4	36.3	310.6
1972	267.2	35.7	302.9
1973	257.6	31.6	289.2
1974	249.9	33.8	283.7
1975	246.3	30.8	277.1
1976	237.8	32.2	270.0
1977	231.0	36.0	267.0
1978	237.5	34.7	272.2
1979	241.7	35.8	277.5
1980	238.4	34.8	273.2
1981	233.2	32.4	265.6
1982	231.2	34.2	265.4
1983	229.3	32.7	262.0
1984[a]	228.4	32.6	261.0

Source: USDA (1985A).
[a] Preliminary.

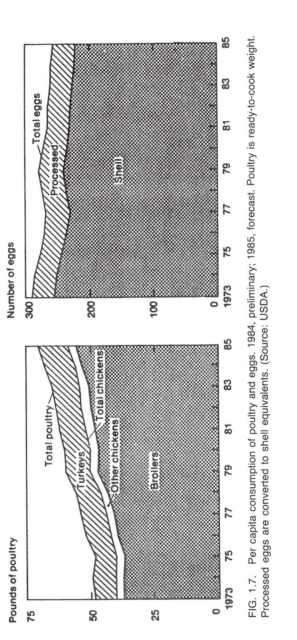

Pounds of poultry

Number of eggs

FIG. 1.7. Per capita consumption of poultry and eggs. 1984, preliminary; 1985, forecast. Poultry is ready-to-cook weight. Processed eggs are converted to shell equivalents. (Source: USDA.)

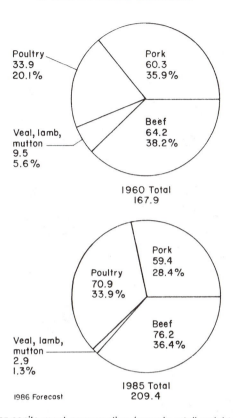

FIG. 1.8. Per capita meat consumption (pounds retail weight equivalent).

growth is the increase in consumption of poultry hot dogs or wieners (Fig. 1.9). In 1977, poultry wieners accounted for 3.2% of total wiener sales. In 3 years they had captured 8% of the market and were selling for 40–50¢ less than those made from pork and beef.

CAREERS IN THE POULTRY INDUSTRY

It appears that the demand for qualified men and women, and especially college graduates, in the poultry industry will continue. Opportunities exist in sales and marketing, scientific laboratories, allied industries, business administration, and field related jobs where one can work directly with birds.

There is also a demand almost worldwide for well-trained, experienced United States Poultry graduates.

To be successful as a poultry producer or technical adviser one must have the following traits.

FIG. 1.9.   Franks from poultry meat now account for over 18% of the market. Chicken franks can be served in a variety of ways to make nutritious family meals.

1. A basic interest and liking for poultry combined with a desire to work with poultry and an understanding of poultry as biological food-producing systems.
2. The ability to supervise, educate, and work with people at all levels of skill.
3. A mind for business, including a willingness to meet contractual obligations.

4. The flexibility to adapt to changes and advancements in science, technology, and management practices within the industry.
5. A realistic and yet altruistic attitude to appreciate one's role as a food producer in an ever-changing food-short world.

REFERENCES

American Poultry Association, Inc. 1983. The American Standard of Perfection. American Poultry Association, Inc., Troy, NY.

Anonymous. 1977A. The Chicken Broiler Industry, Practices and Costs. AE Report No. 381, U.S. Dept. of Agriculture, ERS, Washington, DC.

Anonymous. 1977B. People on the Farm: Broiler Growers. U.S. Dept. of Agriculture, Office of Governmental Affairs, Washington, DC.

Rogers, G. B. Chapter on Poultry and Eggs, pp. 148–189. U.S. Dept. of Agriculture, Agricultural Economic Report No. 441, Washington, DC.

Schertz, L. P., ed. 1979. Another Revolution in U.S. Farming? U.S. Dept. of Agriculture, Washington, DC.

Skinner, J. L. 1978. Chicken Breeds and Varieties. University of Wisconsin Extension, Madison, WI.

USDA. 1985A. Agricultural Statistics. U.S. Government Printing Office, Washington, DC.

USDA. 1985B. Foreign Agriculture Circular, Dairy, Livestock and Poultry Division FL & P- 1 84, Washington, DC.

2

Anatomy and Structure of the Fowl

In the evolutionary scale of animal development, chickens are between mammals and reptiles. In addition to their morphological differences, mammals, birds, and reptiles vary greatly in their methods of reproduction and caring for their young.

In mammals, the young are carried inside the body and nurtured by the mother. Although they are fully developed at birth, they are still dependent on the mother for milk.

Birds lay eggs that contain all the nutrients needed to develop, grow, and sustain the embryo to hatching time with enough extra nutrients to maintain the newly hatched bird for several days after hatching. The parents, usually the female, must keep the eggs warm and protected until hatching. Immediately after hatching, precocial young, while able to feed and run about, are still dependent on the parents for artificial brooding for warmth and protection.

Most reptiles also lay eggs, but the female buries them in sand, soil, or vegetable matter, where the eggs incubate at environmental temperature and hatch. The young must take care of themselves as soon as they hatch since they seldom see their parents.

Unlike reptiles, which are cold blooded, i.e., poikilotherms, birds and mammals are warm blooded, i.e., homeotherms. Although birds have scales on their shanks and toes as do reptiles, they also have feathers and can fly. In order to fly, birds in addition to feathers have compact bodies, rapid digestion, an active metabolic system, and light skeletons, and their musculature, including breast, wing, and leg muscles are well developed to run and fly. Birds and reptiles for the most part are omnivorous and have monogastric digestive systems.

FEATHERS

As birds evolved from reptiles, some of their scales gradually changed and developed into feathers. Chemically, both feathers and scales are composed primarily of a type of protein called keratin. Most animals do not have enzymes to digest keratin, but when feathers and scales are heated under pressure they break down into simpler, digestible proteins that can be used for livestock feed. Depending on their size, birds vary in the number of feathers they have. For example, songbirds have about 2500 feathers, but a swan may have over 10,000.

Birds use feathers for a number of functions including flight and protection from physical injury, temperature extremes, and rain, snow, and wind. The feathers are coated with a waxlike material which the bird gets from its preen gland and spreads over its feathers with its beak. Waterfowl have particularly well-preened feathers, which prevents them from sinking when in the water. Birds also use feathers for camouflage and in courtship behavior.

Feathers grow from the several body regions in arranged patterns or tracts called pterylae. Feathers are classified according to their structure into three groups: contour feathers, plumes, and filoplumes. Contour feathers cover the body and contain most of the feather pigments, for example, lipochrome and melanin are responsible for color and plumage patterns. Plumules or down-type feathers are located under the contour feathers, and serve primarily as an insulation layer. Filoplumes are rudimentary hairlike feathers located mainly in the head and neck region. These hairlike feathers are removed by singeing during processing.

The parts of the feather (Fig. 2.1) include the quill or base and the shaft or rachis, which is a continuation of the quill through the body of the feather. On the underside of the quill is a small downy feather-like structure attached to the shaft, which is called the accessory plume. The main body and most visible portion of the feather is called the web. Barbs project from both sides of the shaft through the web. Barbules branch off the barbs. The barbicels, which are outgrowths off the

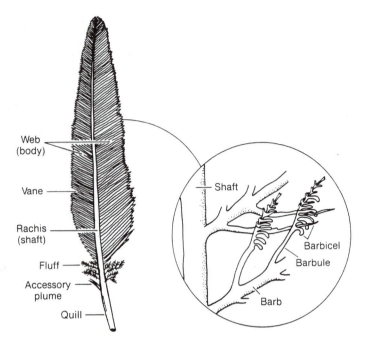

FIG. 2.1. Drawing of a chicken feather.

barbules, contain hooklike projections that lock together much like the teeth in a zipper to give the feather added strength.

Adult birds renew their plumage annually. Called molting the process requires about 6 weeks for all the old plumage to be lost and replaced by new plumage. The old feathers are lost gradually in a regular pattern so the bird has some protection and most can fly during molting; however, waterfowl lose the ability to fly during the molt and regrowth of the primaries.

SKIN

The skin of birds is relatively thin and, unlike mammals, contains no secretory glands. Birds have no sebaceous (sweat) glands; the only exception is the uropygial or preen gland located at the top of the base of the tail which secretes the waxlike material the bird uses to preen its feathers. Birds have a covering of fat immediately under the skin that gives the skin a creamy white or yellow appearance. Carcasses of birds with little fat have a reddish skin.

Skin color is influenced mainly by the subcutaneous fat color, which is determined by genetics and feedstuff. For example, there are breeds of white-skinned (English class) and yellow-skinned (American class) chickens. A group of fat-soluble, yellow pigments called xanthophylls can also be used to produce chickens with a yellow skin. Common sources of xanthophylls include alfalfa meal, marigold petals, and corn gluten meals as natural fat pigmenters.

MUSCLES

Birds have the same three general types of muscle tissue as mammals. Smooth or involuntary muscles are found in blood vessels and the organs making up the respiratory, digestive, and reproductive tracts. These involuntary muscles operate independently and without conscious control by the organism. Cardiac muscle is a specialized type of muscle found only in the heart. Skeletal or voluntary muscles, so called because the organism can control the muscles and use them to move its body and appendages at will, make up the largest mass of skeletal muscles. The skeletal muscles and giblets (heart, liver, and gizzard) muscle tissue are the source of the edible portion of birds that we call meat. For example, the pectoral or breast muscles which are used by birds to fly make up about 50% by weight of the bird's muscle tissue.

Muscles can also be classified as light and dark in color. As food they are referred to as the white meat and the dark meat. Dark muscles contain more of the muscle compound myoglobin, which is rich in iron and used for transporting oxygen. Dark muscles are used for sustained activity such as moving about, walking, or running. Light muscles are associated with flight and for this reason are designed to provide large bursts of energy quickly so the bird can fly. The location of the light and dark meat varies with the function and activity level of the major muscles of each species.

Birds differ from mammals in that they have little interstitial or intermuscular fat usually referred to as marbling in red meat animals. Fat deposition in poultry meat occurs subcutaneously or between muscles accounting for the juiciness found in thigh and leg meat.

SKELETON

The skeletons of most birds are light weight, compact, and adapted for flying. Avian skeletons differ from mammalian skeletons in that they have fewer bones and more bones are fused together for rigidity (Fig. 2.2). For example, part of the vertebrae is fused together to provide the skeleton with the extra strength necessary to absorb the

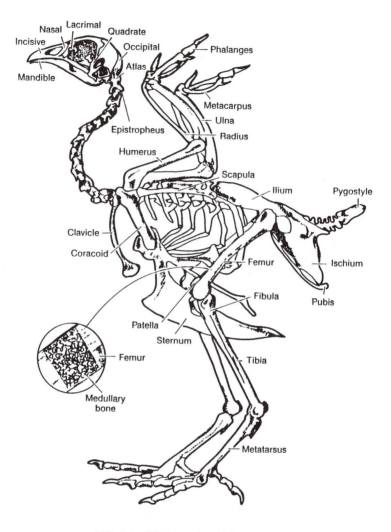

FIG. 2.2.   Diagram of a chicken skeleton.

shocks of take-offs and landings. Birds also possess a pectoral girdle that corresponds to the shoulder girdle of mammals and a pelvic girdle that is open so the female can pass eggs out through the reproductive tract.

Growing or developing bones start as flexible cartilage that becomes hardened as the bones are calcified or ossified from deposition of inorganic salts, particularly calcium phosphate. In long bones, ossification starts in the middle of the bones and gradually extends to the end portion of epiphysis.

Birds also have hollow air-filled pneumatic bones that connect to the respiratory system through openings called foramen. Depending on the species they usually include the sternum, humerus, skull, clavicle, lumbar, and sacral vertebrae.

Since the female of the species lays eggs she must be able to supply calcium for shells as well as calcium for her own maintenance needs. To accomplish this, as females approach sexual maturity and estrogen levels elevate, a unique type of bone tissue, characterized by a honeycombed lacing of bone specules, called the medullary bone develops in the narrow cavity of the femur and to a lesser extent in the tibia, sternum, ribs, and scapula to provide supplemental calcium for egg production. Mobilization of medullary bone tissue occurs when blood calcium levels drop, such as during the night when available calcium absorbed from the gut falls below what is needed in the uterus for shell calcification. When calcium intake from the gut exceeds what is needed to maintain blood calcium levels, calcium is again stored in the medullary bone. Without this control mechanism the hen would be able to store only enough calcium for a few eggs.

CIRCULATORY SYSTEM

The circulatory system in birds includes a four-chambered heart, arteries, veins, capillaries, and blood. The heart is located in the thorax, a little to the left of the median line. As the heart contracts blood is carried from the heart through the arteries under pressure (systolic pressure) and returns through the veins (diastolic pressure); at rest the heart fills as blood returns for recirculation. Blood pressures vary greatly among birds of the same breed, sex, age, and physical condition; for example, an adult male leghorn chicken should have a systolic pressure of 190 and a diastolic pressure of 160 mmHg. Heart rate in general is inversely related to the size of the bird. A small mature chicken's heart generally beats about 300 times a minute while a large turkey tom's heart beats 160 times a minute.

Blood consists of plasma, which is the liquid portion, and cells, which make up about 30–40% of the blood volume. In mature chickens blood amounts to 6 to 8% of body weight while in young growing birds it approaches 10%. The red cells or erythrocytes contain the oxygen-carrying iron pigment hemoglobin. Avian red blood cells (erythrocytes) are nucleated whereas mammalian cells are not.

Erythrocytes, leukocytes (white cells), and thrombocytes (analogous to platelets in mammals) are formed primarily by the spleen and bone marrow which also serve as storage sites. Mature chickens normally have 2.5–3.5 million cells/mm^3. Leukocytes function primarily to aid the organism in preventing disease, whereas thrombocytes are involved with blood clotting.

BODY TEMPERATURE

Birds are able to maintain a fairly constant deep-body temperature, meaning they are homeotherms. They can also increase their body temperature by generating heat within the tissue as opposed to heat gained from the surroundings, meaning they are also endotherms. The deep body temperature of birds fluctuates during the course of the day; this phenomenon is known as the circadian rhythm. The daily cycle of body temperature is keyed to the photoperiod. Body temperature tends to be highest when the bird is most active. Body temperatures of chickens fluctuate between 105 and 109°F (41–43°C) with a deep body temperature of 41.5°C. Lethal temperatures for chickens appear to be about 117°F (47°C) depending on humidity, age, and other related factors.

RESPIRATION

The structure and functioning of the avian respiratory system are different from the mammalian system. Expiration is the active part of respiration, which is the opposite of mammals. The lungs of birds are rather small rigid structures and are attached to the skeleton in the thoracic cavity. Avian lungs expand or contract little during respiration and there is no true diaphragm. In addition to the lungs, birds have nine air sacs, four paired sacs, and a single median sac located in the thorax. The air sacs connect the lungs to the pneumatic bones, increase buoyancy, and help impart lightness to the bird for flight. Since air sacs are transparent, delicate membranous structures, they can be easily irritated by dust and ammonia in poorly ventilated poultry houses.

The rate of respiration is controlled in the medulla of the brain by the carbon dioxide rather than the oxygen level. As carbon dioxide levels increase the respiration rate increases. The respiration rate in an adult bird such as a chicken ranges from 20 to 35 expirations per minute depending on sex and size.

Basically, respiration in birds is accomplished in the following manner (Fig. 2.3). The abdominal muscles relax dropping the rear of the sternum. Then air flows in through the lungs and air sacs. When abdominal contractions occur, the rear of the sternum is drawn up and air is forced out of the air sacs through the lungs.

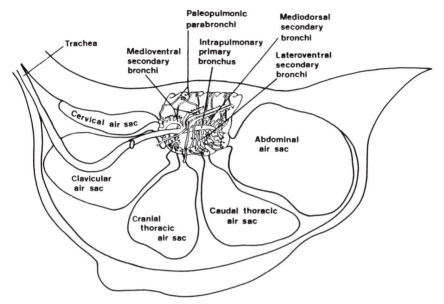

FIG. 2.3. Drawing of the avian respiratory system.

DIGESTION

Birds have a relatively simple monogastric digestive system (Fig. 2.4), somewhat similar to that of a pig or man. There is little time or place for microorganisms to help digest food, as in ruminants. Birds, like humans, must depend on enzyme secretions to aid in breaking down complex molecules to simpler nutrients that they can absorb. Their digestive system is composed of the alimentary tract, liver, and pancreas. These organs serve the same functions as in mammals, i.e., intake of food, storage, digestion, and elimination of body wastes.

MOUTH AND ESOPHAGUS

Birds have no lips or teeth. The beak or mandibles are used in picking up food and breaking it up to a size that can be swallowed. For this reason, the size and shape of the mandibles are a good indicator of a bird's dietary habits. For example, the tongues of many seed- and insect-eating birds, such as the chicken, are arrowhead shaped and hinged to force food back into the esophagus. Most birds, such as the chicken, have a hard palate. To swallow they elevate their head to cause a pressure differential in the esophagus. Salivary glands secrete mucous saliva to coat and lubricate food particles for easy swallowing.

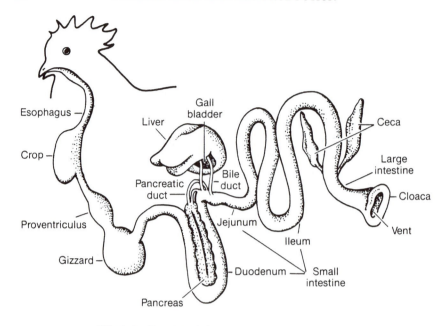

FIG. 2.4.   Drawing of the digestive tract of a chicken.

CROP AND PROVENTRICULUS

The crop serves as a temporary storage area in the esophagus. Food is softened there by the moisture of ingested water and saliva. Waterfowl have an expandable esophagus instead of a pouch-like crop. Chickens and turkeys have pouch-like crops whereas many small birds have no crops.

The proventriculus or glandular stomach of a bird is a relatively small thickening of secretory tissue at the end of the esophagus. It secretes an enzyme, pepsin, which aids in protein digestion and hydrochloric acid, which initiates breakdown of seed coats and alters the pH of the food to about 2.5 in chickens. Little, if any, digestion of food takes place in the proventriculus.

GIZZARD

The oval muscular stomach or gizzard of the bird was also found in some prehistoric reptiles. It consists of two pairs of thick powerful muscles located between the proventriculus and duodenum. The gizzard breaks down ingested feeds into smaller more digestible fractions. Birds normally consume some gravel or grit to abrasively aid the physical breakdown of feed particles. The muscles of the gizzard are protected from the acid pH by a thick epithelial layer. Muscle contractions of the gizzard are strong enough to bend metal objects such as nails that may be swallowed. Since modern-day poultry rations are ground before feeding, the gizzard is not as necessary for good digestion as when whole grains are consumed. This is why gizzards of modern-day chickens are relatively tender.

PANCREAS

Below the gizzard is a folded loop in the small intestine known as the duodenum. In the middle of this loop is the pancreas, which secretes enzymes capable of hydrolyzing protein, carbohydrates, and fats. The pancreas also secretes insulin to regulate sugar metabolism.

LIVER

The liver is made up of three large lobes of brownish soft tissue. The bile, a greenish-clear alkaline liquid, is secreted in the liver and stored in an enlarged area of the bile duct under the right lobe of the liver known as the gallbladder. Food entering the duodenum causes the gallbladder to contract and secrete bile into the intestine. The main function of bile is to emulsify fats so they will enter into solution to be digested. pH control is further carried out by pancreatic juice. The bile duct from the left lobe directly enters the small intestine. Unhealthy birds with gastroenteritis are off feed and fecal materials are tinged green from the bile secreted into the small intestine.

SMALL INTESTINE

The small intestine consists of the duodenal loop and the lower portion of the small intestine. In an adult chicken it is almost 5 ft in length. Enzymes secreted by the pancreas, including amylase, lipase, and protease, help to continue the breakdown of starches, fats, and proteins. Additional enzyme secretions from the walls of the intestine break down peptides to amino acids and complex sugars to simpler ones that can be absorbed (Table 2.1). No specialized area exists for bacterial action to alter foodstuffs in preparation for digestion. Diges-

TABLE 2.1. Poultry Enzymes and Digestive Juices

Enzyme	Digestive juice in which enzyme is found	Digestive juice secreted by	Part of digestive tract into which digestive juice is secreted	Enzyme acts on	Enzyme active in acid or alkaline medium	Action produces
Salivary amylase	Saliva	Salivary glands	Mouth	Starch	Neutral to slightly alkaline	Maltose
Salivary maltase	Saliva	Salivary glands	Mouth	Maltose	Neutral to slightly alkaline	Glucose
Pepsin	Gastric juice	Wall of proventriculus	Proventriculus	Proteins	Acid	Proteoses, polypeptides, peptides
Gastric lipase	Gastric	Wall of proventriculus	Proventriculus	Fats	Acid	Fatty acids, glycerol
Pancreatic amylase	Pancreatic juice	Pancreas	Upper small intestine (duodenum)	Starch	Alkaline [a]	Maltose
Trypsin, etc.	Pancreatic juice	Pancreas	Upper small intestine (duodenum)	Proteins, proteoses, polypeptides, peptides	Alkaline [a]	Intermediate protein breakdown products, amino acids
Pancreatic lipase	Pancreatic juice	Pancreas	Upper small intestine (duodenum)	Fats	Alkaline [a]	Fatty acids, glycerol, mono-glycerides
Intestinal peptidases (erepsin)	Intestinal juice	Wall of small intestine	Small intestine	Intermediate protein breakdown products	Alkaline [a]	Amino acids
Intestinal maltase	Intestinal juice	Wall of small intestine	Small intestine	Maltose	Alkaline [a]	Glucose
Sucrase	Intestinal juice	Wall of small intestine	Small intestine	Sucrose	Alkaline [a]	Glucose
Lactose	Intestinal juice	Wall of small intestine	Small intestine	Lactose	Alkaline [a]	Glucose, fructose
	Bile	Liver	Upper small intestine (duodenum)	Bile reacts with fats	Alkaline [a]	Soap, glycerol

Source: Patrick and Schaible (1980).
[a] pH ranges from 3 to 6.8 in poultry.

tion and absorption take place rapidly in the small intestines of birds, often in as little time as 3 hr or less in fasted poultry. The surface area for absorption is high in birds allowing the intestine to absorb nutrients rapidly.

CECA

Ceca are blind pouches, located at the intersection of the small and large intestine. They approach 6 in. in length and extend forward from the junction of the intestine. Presumably, at one time in the evolutionary process, the ceca was able to digest fiber by microbial action. Today adult chickens even on high fiber diets show little evidence of fiber digestion.

LARGE INTESTINE, RECTUM, AND CLOACA

The large intestine is short (4–5 in.) and extends from the small intestine to the cloaca. This portion is commonly called the rectum. The cloaca, in addition to holding fecal material, is a common passage for the urinary and reproductive tracts. The urinary tract empties into the cloaca and is excreted with the feces, as a solid chalky material consisting mainly of uric acid, through the anus or vent of the bird. The urine of birds is the solid white material on their droppings.

EXCRETORY SYSTEM

The excretory system consists of the elongated three-lobed kidneys and the ureters (Fig. 2.5). The ureters carry waste in the form of uric acid from the kidneys to the cloaca, for passage from the body. Uric acid is a water-insoluble compound, occurring as an end product of nitrogen metabolism. The functional units of the kidney are the collecting tubules or nephrons.

NERVOUS SYSTEM

The nervous system includes the brain and spinal cord, with branches leading to the sensory organs and the sympathetic nerves, controlling the viscera. The body functions are controlled by stimuli from the sense organs, through responses from the nervous system. The voluntary actions of the body are controlled by the cerebrospinal por-

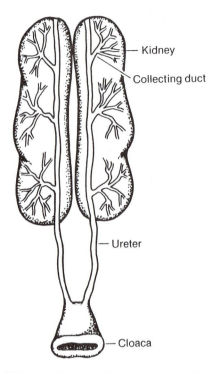

FIG. 2.5. Anatomy of the excretory system.

tion of the central nervous system. The involuntary actions of organs, vascular systems, and glands are controlled by the autonomic system.

The cerebral cortex is small in birds. The hypothalamus is well developed and influences aggressive and sexual behavior, feed and water intake regulation, and regulation of anterior pituitary secretions.

AVIAN SENSES

Birds have a thin membrane forming a third eyelid (nictitating membrane) that serves to protect the eye. In birds the orbits of the eye are very large compared to mammals suggesting that birds are vision oriented. Many birds, including the chicken, have binocular as well as monocular vision. The location of the eyes in the skull varies among bird species, but many birds, such as the chicken, have fields of vision over 300°. The visual acuity of chickens and many birds is probably no better than humans. However, the rapid reaction time of birds to visual stimuli makes them appear to have keen eyesight.

Birds have a reptilian-type ear and well-developed hearing. To survive they must react quickly to sounds they hear or become victims of predators. The sense of smell in birds is less highly developed and though olfactory information is obtained, it is not often behaviorally meaningful in species such as poultry.

Compared to other animals the chicken has only 24 taste buds. For example, pigeons have 37, Japanese quail 62, humans 9000, and catfish 100,000.

Birds are indifferent to sweets and tolerate relatively bitter foods. Terrestrial birds such as the chicken have a low tolerance for salt in their water. Young chicks will refuse a toxic 2% salt solution. Many marine and aquatic birds are equipped with a salt gland, located near the orbit of the eye at the top of the beak, which filters salt from ingested fluids and discharges it through the nares in a concentrated salt solution.

ENDOCRINE ORGANS

Endocrine organs are important in controlling body processes. In birds endocrine glands or organs include the pituitary (hypophysis), the adrenals, the thyroid, the parathyroids, the pancreas, the gonads (testes and ovary), the ultimobranchial glands, the kidneys, and the intestine. The thymus and pineal bodies are sometimes classified as endocrine organs but their functions are not clearly understood. Endocrine secretions are carried to the site of action by the blood stream where they act through chemical stimulators.

The anterior pituitary (adenohypophysis) produces hormones involved in the regulation of other endocrine organs. Hormones produced by the anterior pituitary include the adrenocorticotropic hormone (ACTH), melanotropin (MSH), the growth hormone (GH), the thyrotropic hormone (TSH), as well as the gonadotropic hormones (follicle-stimulating hormone, luteinizing hormone, and prolactin).

The hypothalamus produces releasing factors that control the release of regulating hormones from the anterior pituitary. Sensory receptors in the nervous system receive stimuli and cause the release of hormone-releasing factors in the hypothalamus which can cause the further release of another hormone. Light effects on reproduction in birds is a good example of this interaction.

Adrenal secretions aid the bird in coping with stress. The metabolic rate and feather growth are influenced by the thyroid hormone. Calcium deposition as well as mobilization are controlled by secretions from the parathyroid and ultimobranchial body. The kidneys secrete several hormones that are primarily involved with formation of blood cells and regulation of blood pressure. Hormones involved in reproduction will be discussed in the next chapter.

REFERENCES

Limberg, P. R. 1975. Chickens, Chickens, Chickens, 1st Edition. Thomas Nelson, Nashville, TN.

Nesheim, M. C., Austic, R. E. and Card, L. E. 1982. Poultry Production, 12th Edition. Lea & Febiger, Philadelphia, PA.

Patrick, H., and Schaible, P. J. 1980. Poultry Feeds and Nutrition, 2nd Edition. AVI Publishing Company, Westport, CT.

Sturkie, P. D. 1976. Avian Physiology, 3rd Edition. Springer-Verlag, NY.

Winter, A. R., and Funk, E. M. 1941. Poultry Science and Practice. Lippincott, Chicago, Philadelphia, New York.

3

Physiology and Reproduction in Poultry

ENDOCRINE SYSTEM

Endocrine glands are ductless glands that secrete hormones directly into the bloodstream to bring about a number of important physiological functions. They include the ovary, testes, pineal, thyroid, parathyroids, ultimobranchial gland, adrenal, pituitary, and islets of Langerhans of the pancreas. In addition, hormones are also secreted in the brain and gastrointestinal tract (Fig. 3.1).

The anterior pituitary produces secretions that effectively regulate the secretion of other endocrine glands, including thyroid-stimulating hormone, adrenocorticotropic hormone, gonadotropic hormone, and growth hormone.

Carbohydrate and mineral metabolism as well as the bird's ability to handle stress are affected by the adrenal gland. The secretion of digestive juices is handled by hormones from the gastrointestinal tract, which may influence passage of food and contraction of the gallbladder. Carbohydrate metabolism is regulated by insulin and glucagon

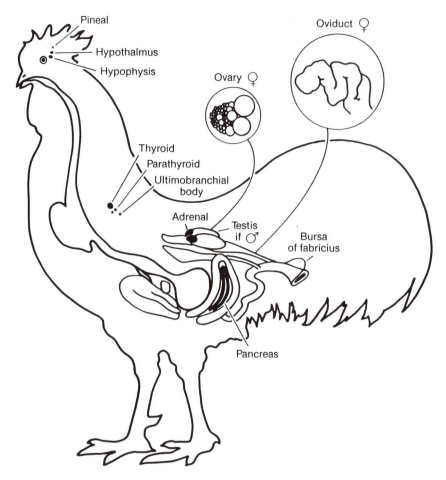

FIG. 3.1.   The location of endocrine and reproductive organs in the chicken.

which are produced by the islets of Langerhans and the beta cells of the pancreas.

The metabolic rate of the bird is controlled by thyroxin produced in the thyroid. Thyroxin also influences feather growth and color. Parathyroid hormone secreted by the parathyroid glands helps regulate calcium and phosphorus levels and is involved in egg shell formation and the mobilization of medullary bone. Calcitonin produced by the ultimobranchial gland is also involved in calcium regulation and deposition.

Posterior pituitary hormones aid in water balance and regulation of blood pressure. Another posterior pituitary hormone, oxytocin, is involved in actual egg laying (oviposition).

PHYSIOLOGY OF REPRODUCTION

Reproduction in healthy mature male and female avian species occurs through the systematic integration of many body processes. Egg laying, ovulation, and the maturation of spermatozoa are probably the three most important single factors involved in reproduction. They are influenced by the effect of light and darkness, as well as the action of the nervous and endocrine systems. Many birds, including the species used for poultry production, are sensitive to photoperiod. This is the reason most common bird species mate and rear their young in the springtime as day length increases, not because of the onset of warm weather. Mating activity as well as egg laying stop as day length decreases and autumn approaches. Contrary to popular belief, egg laying ceases because of changes in day length, not because of the onset of winter. Domestic poultry respond to changes in day length in a similar way.

Today, the use of light as a management tool is essential in modern-day poultry production to stimulate egg production on a year-round basis. Length of the photoperiod rather than intensity of light is the critical factor in stimulating egg production. Intensity levels of 0.5 ft candle (5 lux) at bird eye level are sufficient to stimulate egg production. In housing systems using no natural light, levels of intensity of above 1.0 ft candle (10 lux) are unnecessary and costly.

Birds have reptilian-type eyes and do not respond to all wavelengths of light. They seem to be stimulated by wavelengths from 6640 to 7400 Å. Figure 3.2 illustrates the visual wavelengths for humans and chickens. Shorter wavelengths do not stimulate reproduction. Incandescent bulbs give off effective wavelengths of light, but fluorescent bulbs are effective only if the wavelengths are in the range of 6640–7400 Å. Figure 3.3 shows the wavelengths contained in the sun, incandescent, and fluorescent lights.

When artificial light is used to stimulate poultry, several details need to be rigidly followed. Hens in production cannot be subjected to decreasing day length or they will go out of production. Scientists have developed many satisfactory variations of lighting programs; however, the chances of human error or mechanical failure warrant careful consideration before adaptation of these programs. Fourteen hours of light stimulates egg production in most poultry species. A popular lighting program frequently used for commercial layers consists of 14 hr of light from 22 weeks of age through the peak of egg production. Then as production starts to decline, the light is increased to 16 hr per day and continued at this light level until the end of the laying period.

When egg laying starts at any early age before physical maturity of the pullet, egg size is small and the eggs sell for less. This can be a

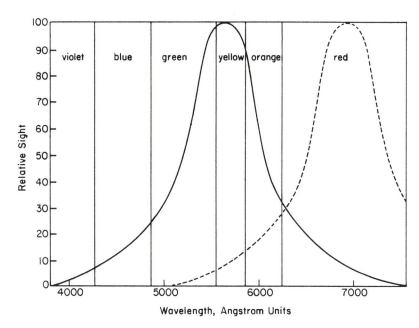

FIG. 3.2.   Visual wavelengths that affect humans and birds. Solid curve, human eye sensitivity; dashed curve, chicken eye sensitivity.

problem with winter-hatched pullets that mature under natural light with increasing daylength. Pullets are produced continually throughout the year but are reared in facilities in which the grower has control of the light and after about 6 weeks of age restricts the amount the pullets receive to 6–8 hr a day to prevent early sexual maturity and the laying of small eggs.

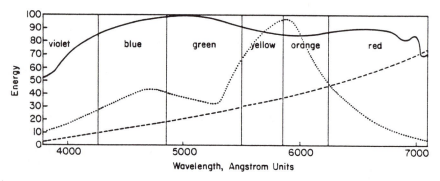

FIG. 3.3.   The electromagnetic spectrum. Standard voltage equals the slope of the incandescent curve. Solid curve, sunlight; dashed curve, incandescent light; dotted curve, cool fluorescent light.

Secretion of hormones associated with reproduction is influenced by light in the following sequence. Photoreceptors in the hypothalamic region of the brain receive light and these sensory receptors stimulate the nervous system which causes the release in the hypothalamus of hormone-releasing factors that cause other endocrine glands to secrete hormones. One current theory is that the nervous system helps regulate body functions and sends messages by electrical stimuli while the endocrine system controls body processes. The messages from the endocrine system are chemical stimulators carried by the bloodstream to the site of action.

REPRODUCTION

Reproductive organs in birds include the testes in the male and the ovary and oviduct in the female. During embryo development two ovaries and oviducts are present in avian females. The left ovary and oviduct become functional in most bird species and all poultry. Some raptors and other species have functional right ovaries and oviducts instead of the left ovary. The reason for this is not known, but it also occurs in some fish, reptiles, and monatremes, so it obviously is not a flight adaption. Figure 3.4 illustrates the female avian reproductive system.

Reproduction in the Female

The Ovary   The left ovary is located in the avian female on the left side of the body (right side as you postmortem) at the cephalic end of the kidneys. The ovary is attached to the body wall by the mesovarian ligament.

A mass of 2000–12,000 or more small ova make up the ovary of an immature pullet. As a result modern-day production practices, probably only 200–300 of these are ever ovulated in chickens. Obviously, the number ovulated is fewer in other birds. Ovarian tissues are a pale yellow and are similar in color to testicular tissue.

Immature avian females possess a small undeveloped ovary and oviduct. As sexual maturity approaches, the developing ovary secretes estrogen. The oviduct then develops under the influence of estrogen as well as the initiation of medullary bone formation. Increases also occur in blood calcium, protein, fat, and vitamin levels necessary for formation of egg components. Ovarian follicle development is initiated by gonadotropic hormones. The ovary reaches maturity, starts secreting estrogen, and from 7 to 10 tiny ova on the ovary go into a rapid stage of

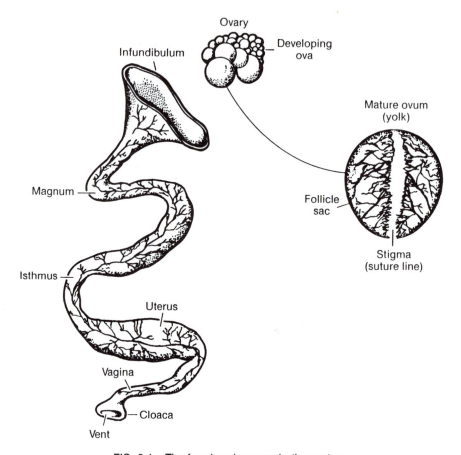

FIG. 3.4. The female avian reproductive system.

development. During the approximately 10 days it might take for the first ovum in the rapid growth stage to mature, the pubic bones soften and spread and the vent enlarges in preparation for oviposition. Another hormone produced by the ovary is progesterone which stimulates the hormone-releasing factors of the hypothalamus to release luteinizing hormone (LH) from the anterior pituitary. LH is responsible for the ovulation of a mature yolk from the ovary. Oviduct function is also influenced by progesterone. Actual ovulation occurs when the follicle sac containing the mature yolk ruptures and tears down the less vascular portion known as the stigma or suture line. The mature ovum or yolk is complete at this point. It is engulfed by the infundibulum shortly after ovulation. A blood spot can occur in an egg since during ovulation a slightly irregular tear of the stigma into the more vascular area can cause the formation of a drop of blood on the yolk and subsequently in the shell egg. In our current egg-grading standards, eggs with blood spots are not used for human food because of their unfavorable aesthetic value, not because they are not as nutritious and wholesome as other eggs. The actual expulsion of the egg after it passes through the oviduct is influenced by arginine vasotocin released from the posterior pituitary gland and local tissue hormones called prostaglandins.

Some of the male sex hormone, testosterone, is secreted by the ovary under the influence of gonadotropic hormones of the pituitary and results in the red waxy comb and wattles of the hen. Testosterone may also affect the secretion of albumen.

The Oviduct   The oviduct of the avian female is a relatively large folded tube of tissue occupying a majority of the space on the left side of the abdominal cavity. The oviduct consists of five distinct parts which vary in size depending on the state of reproduction. The principal parts of the oviduct are as follows with the approximate length listed for a 4-lb hen: (1) the infundibulum or funnel (9 cm), (2) the magnum (33 cm), (3) the isthmus (10 cm), (4) the uterus or shell gland (10 cm), and (5) the vagina, the portion leading from the uterus to the cloaca.

After ovulation takes place, the yolk or ovum is picked up by the infundibulum normally in less than 20 min. The action of the infundibulum is conditioned by ovulation. The infundibulum also contains the infundibular sperm storage site which may represent a short-term or transient place of residence in which sperm gather just before fertilization. Occasionally, the yolk is not picked up in the infundibulum and is reabsorbed by the hen in as little as 24 hr. Fertilization of the avian egg occurs after ovulation and before albumen secretion begins; therefore it occurs while in the infundibulum or as it enters the magnum.

The albumen or egg white is secreted in the magnum around the yolk or ovum in layers. The chalaziferous layer is attached to the yolk and is twisted clockwise at the large end and counterclockwise at the small end. This implies that the action of the ovum through the magnum is a mechanical one as it gathers albumen from the glandular cells of the magnum. The chalazae are mucin fibers arising from the initial inner layer of albumen. The other layers of albumen include the inner thin albumen, the thick albumen, and the outer thin albumen. Approximately 60% of the albumen is from the thick and 25% from the outer thin layer. The albumen is uniform as it is secreted, but the movement through the oviduct and the addition of water appear responsible for the separation into layers.

Egg albumen consists of a number of different proteins. These include ovalbumen, 54%; conalbumen, 13%, ovomucoid, 11%, ovoglobulin, 37%, lysozyme, 3.5%, and ovomucin, 2%. Ovomucin is responsible for the gel-like quality of egg white.

The albumen-layered ovum moves from the magnum to the isthmus by peristaltic movement where the two shell membranes are added. The two shell membranes consist of the protein keratin and loosely fit the egg contents in the isthmus. The membranes adhere to each other and the outer membrane is approximately three times as thick as the inner membrane. The shell membranes are semipermeable and permit passage of water and crystalloids. The tissue of the isthmus is fairly easy to distinguish from the magnum because the folds of the secretory glands are less distinct and numerous. The isthmus through the shaping of the shell membranes has considerable influence on the shape of the egg. The egg passes through the isthmus and has the shell membranes formed in most hens in about 1.25 hr.

The pouch-like uterus is thick and muscular and contains tubular and unicellular glands. Presumably these cells form the watery uterine fluid used to plump the flaccid egg as it arrives from the isthmus. Water and salts are added to the albumen and pigments are added to the shell. The brown pigment seen in eggshells is porphyrin synthesized from δ-aminolevulinic acid and is uniformly distributed through the shell in chicken eggs. In chickens, this pigment is deposited within the last 5 hr of shell formation. The egg remains in the uterus about 20 hr to allow for adequate calcium deposition to form the shell. The eggshell is about 98% calcium carbonate in the form of calcite. It also contains magnesium, phosphate, and citrate in small amounts, as well as traces of sodium and potassium. The organic matrix of the shell is a glycoprotein that forms a fine fibrous net in the basal caps and the inner parts of the mammillae. The initial calcification of the mammillary cores actually occurs in the isthmus.

The diets of commercial laying chickens may contain as much as 3.6% actual calcium in the diet. Much of the eggshell calcium comes directly from the calcium and food absorbed from the intestine. At night, when little calcium is absorbed from the gut, some calcium is mobilized from the medullary bone and used to maintain blood calcium levels. Nearly 2 g of calcium is deposited in each eggshell, resulting in hens having a higher calcium requirement than even a milk cow. Calcium turnover in laying birds is much more rapid than in mammals. A hen producing 250 eggs a year secretes calcium in the eggshell during the year corresponding to 20 times the weight of the calcium in her skeleton. Adequate shell quality is needed to prevent breakage in handling and marketing eggs; currently, this breakage causes considerable loss to the industry.

The blood plasma of a laying hen contains from 20 to 30 mg of calcium per 100 ml. The formation of an eggshell requires 100–150 mg of calcium/hr. Calcium must be continually replaced through intestinal absorption and mobilization from the medullary bone. When dietary calcium is higher than 3.56% most of the calcium comes directly from the intestine. If the calcium level is 1.95%, the medullary bone supplies 30–40% of the shell calcium. The relationship varies depending on time of day, since most of the shell is formed at night when the calcium content of the digestive tract declines. Bone sources of calcium may be particularly important in early morning hours before the lights come on and the bird feeds again.

Eggs move from the oviduct into the vagina and are expelled through the cloaca. Eggs appear to be formed with the small end down as they move through the oviduct, yet a large proportion of eggs are laid large end first, probably to more effectively exert muscle force for oviposition. If the female is going to lay eggs on consecutive days, ovulation of another ovum will take place within 15–30 min of oviposition. Shell formation in the uterus consists mainly of crystal growth and further matrix deposition and the formation of the cuticle.

The eggshell consists of the inner and outer shell membranes followed by calcium deposits and the formation of the mammillary, pallisade, and crystal layers of the shell. Figure 3.5 is a diagram of the crystalline structure of the eggshell. These layers are penetrated by numerous pores, which, in turn, are filled with cuticle. Cuticle is an organic compound that seals the pores reducing moisture loss and preventing bacterial penetration. Much of the cuticle is removed from table eggs when they are mechanically washed. To replace the cuticle table eggs are sprayed with a light mineral oil mist to aid in preventing spoilage by filling the pores and slowing bacterial penetration.

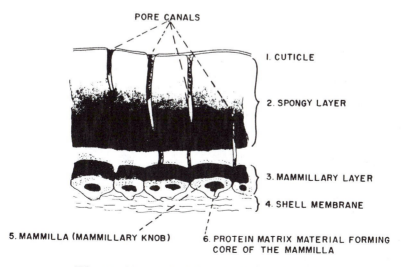

FIG. 3.5.   Magnified radial section through the shell.

Eggshell formation requires adequate calcium in the diet so that the level of calcium ions to the uterus in the presence of carbonate ions in the uterine fluid is adequate to form calcium carbonate. The relationship between blood calcium, CO_2, and bicarbonate ions in the blood and uterus is depicted in Figure 3.6. Carbonic anhydrase is found in the mucosa of the uterus and is responsible for the formation of the bicarbonate ion from CO_2 and H_2O. The acid–base balance of the blood influences shell formation. Excess hydrogen ions in the blood, known as metabolic acidosis, interfere with the calcification process by preventing H^+ ion production when CO_3^{2-} is formed in the uterine fluid. When female birds are laying in hot weather and pant to lose heat by water loss through respiration, CO_2 and HCO_3^- ions are reduced in the blood. The blood buffering capacity is lowered and may result in metabolic alkalosis or may just cause poor buffering of the hydrogen ions produced during shell formation, again interfering with CO_3^{2-} production; this may be the cause of thin shells in hot weather.

Rate of Lay and Laying Cycle   The chicken, duck, and *Coturnix* quail lay a number of eggs on successive days; this is called a sequence. The sequence is then interrupted by a rest period for one or more days before laying continues. Some researchers use cycle or clutch to describe the laying sequence but neither term adequately describes the laying pattern. Some variations in laying patterns are illustrated as follows:

XXXX - XXXX - XXXX	4-egg cycle
XXX - XXX -- XXX --- XXX	regular sequence, irregular skip
X - XX - X - XXXX -	irregular sequence, regular skip
XXXXXX -	long sequence

A good laying hen normally will lay four to six eggs in a sequence. The interval between eggs laid on successive days by most hens ranges from 24 to 28 hr depending on the length of the sequence.

Size, Shape, and Color of Eggs   The characteristic shape of the egg is determined in the magnum but can be modified in the isthmus or the uterus.

The small size of eggs from pullets just beginning to lay results from the smaller yolks in the eggs as well as less albumen secretion. The shell is formed to fit egg contents regardless of size.

The position of an egg in the lay sequence affects its weight. The first egg in a sequence is usually the largest; the size then decreases slightly through the sequence. The decrease can be attributed to a decline in the amount of albumen secreted because yolk weight appears to re-

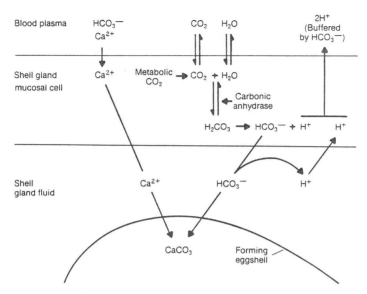

FIG. 3.6.   Carbonate portion of the eggshell.

main constant for all eggs in the sequence. During extended hot weather, egg size declines as feed intake drops and protein intake is lowered.

Double-yolked eggs result from the movement of two yolks through the oviduct at the same time. The double yolks result from simultaneous ovulations or delays in a previous yolk's passage. It appears that the majority of doubled-yolked eggs result from simultaneous ovulations. Many pullets coming into production lay a few double-yolked eggs from simultaneous ovulations as the complex control mechanisms of the ovary, oviduct, and endocrine system become synchronized. Triple-yolked eggs are rare but can occur in some genetic lines rather frequently. Blood spots on the yolk as mentioned earlier occur as the result of an irregular tear on the stigma or suture line of the follicle sac during ovulation.

Eggshell colors for the consumer markets are primarily white with some specific market demands for brown-shelled eggs. Hens laying brown shelled eggs show a gradual lightening of shell pigmentation during the sequence of lay. Occasionally, someone with a small family flock will find blue or green pigmented eggshells. The blue color is a simple dominant trait found in the Aracuna variety of chickens of Central and South America. Once this trait enters a flock, various shell colors appear including blue and various greens when crossed with brown egg layers. These chickens are often advertised as Easter egg chickens but have no advantages over other chickens except various shell colors.

It appears some misshapen eggs occur because an egg remains in the oviduct longer than normal and the next egg in the sequence catches up with the hard-shelled egg in the uterus. The flaccid second egg pushes against the hard-shelled egg distorting the shape as it calcifies or prevents normal calcification on one side making the egg thin shelled. Some avian viral diseases can also alter eggshell shape as well as affect interior quality. Respiratory diseases such as bronchitis can affect albumen quality and Newcastle disease can cause misshapen eggs.

The Complete Egg   The parts of an egg are shown in Fig. 3.7. The small white disk on the surface of the yolk is the germinal disk or blastoderm where embryo development begins in fertile eggs. The germinal disk is present even if the egg is infertile. The yolk material is contained in a thin membrane called the vitelline membrane which gives the yolk its shape. The yolk material is formed in concentric layers of light and dark yolk. A light layer and a dark layer are laid down over a 24-hr period. The lighter, thinner rings indicate that less pigment was available, as would occur during the early morning hours after most nutrients were absorbed from the digestive system. The darker, thicker rings indicate that absorption of nutrients was higher during the day as the bird feeds. The latebra acts as a stalk keeping the germinal disk floating upright and as a stabilizer of the concentric rings. Carotenoid pigments are responsible for yolk color and consist mainly of cryptoxanthine, alcohol-soluble xanthophylls, and carotenes. Vitamin A is also included in this group.

The albumen or egg white is a thin gelatinous material consisting of 88% water and about 11% protein. Surrounding the albumen are the proteinaceous shell membranes. When an egg is laid it is at the body

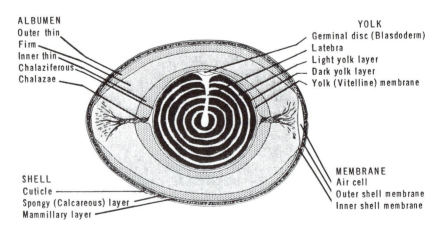

FIG. 3.7.   Structure of a hen's egg.

temperature of the hen; since the ambient temperature is lower than the body temperature, egg contents cool and contract as the egg loses gases and moisture. The two shell membranes separate generally at the large end of the egg with the other membrane sticking to the shell and the inner membrane to egg contents forming the air cell.

The eggshell is primarily composed of calcium carbonate with lesser amounts of magnesium carbonate and calcium phosphate. Organic matter makes up less than 4% of the shell. The shell consists of about 11% of the total egg weight and has as many as 8000 pores per shell. These pores are arranged so that the majority are above the equator of the egg toward the large end. The shells of the eggs of domestic chickens vary in thickness but should be about 15 thousandths of an inch thick. The average weight of an egg is considered to be in the range of 2 oz or 60 g.

Reproduction in the Male

The reproductive tract of male birds is quite simple compared to mammals. It consists of the paired testes, the epididymi, the vas deferentia, and the rudimentary penis (Fig. 3.8). The testes of birds are located near the cephalic end of the kidneys and are ventral to them. Testes of birds are retained in the body cavity and never descend into a scrotum. They are light yellow in color, rather ellipsoid in shape, and have a well-defined surface vascular system. In mature chicken males, the testes are approximately 1% of body weight. Avian species have no Cowper's or prostate gland. The testis consists of seminiferous and rete tubules and vas efferentia. The penis of gallinaceous birds is small and when erected is engorged with lymph. Waterfowl have well-developed phalli that are spirally twisted.

The sperm pass from the seminiferous to the rete tubules then to the vas efferentia, epididymis, and vas deferens. Sperm are normally stored in the vas deferens.

The spermatozoa of avian species are longer than mammalian, but the head is narrower so the spermatozoa are torpedo shaped rather than paddle shaped (Fig. 3.9). Avian spermatozoa have been described as having a simple acrosome with the midpiece being a cylindrical distal centriole surrounded by a cover of mitochondria. The chemical and physical properties of avian semen differ from mammals probably because of the absence of seminal vesicles and prostate glands. The seminal plasma has little fructose, citrate, inositol, phosphatidylcholine, ergothioneine, and glyceryl phosphorylcholine. Potassium and glutamate levels are high, while chloride levels are low. Avian semen varies in color from male to male as the concentration of spermatozoa

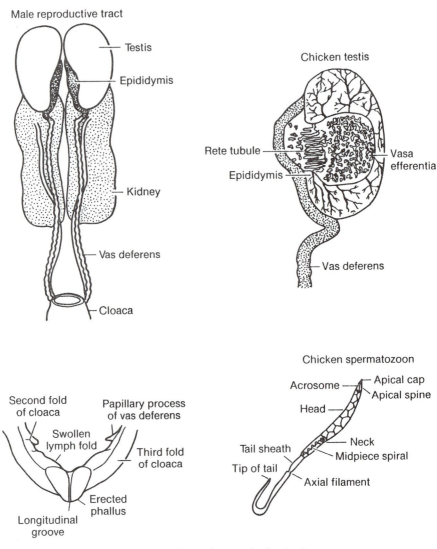

FIG. 3.8.   The male reproductive tract.

varies, but it is generally white and opaque with a pH in the chicken of 7.0–7.6.

Generally in chickens, the volume of semen collected from trained males would be from 0.5 to 1.0 ml volume per ejaculate with a concentration of about 1.7 to 3.5 billion spermatozoa. Turkeys generally produce less volume of semen, about 0.2 ml, but the concentration of spermatozoa is higher.

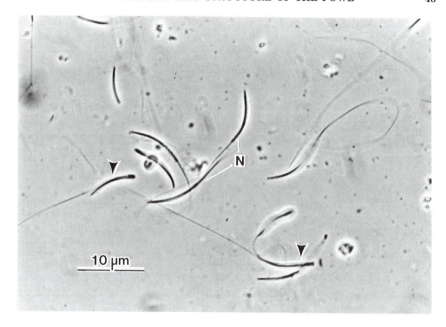

A

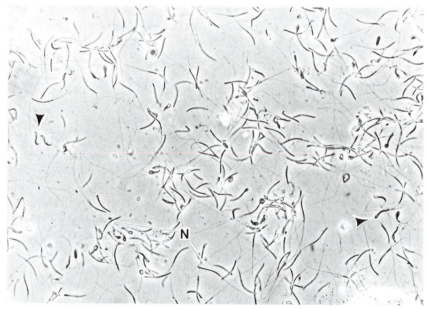

B

FIG. 3.9.   (A) Morphologically normal (N) and abnormal spermatozoa (arrowheads). The most prevalent anomaly is bent sperm in which the sperm head folds backward and lies parallel and adjacent to the sperm tail. Bent sperm remain highly motile but, as their numbers increase, hen fertility decreases. (B) Morphologically normal (N) and abnormal spermatozoa (arrowheads). Among the anomalies are bent (arrowheads) and coiled (doughnut shape) spermatozoa. (Source: USDA.)

Natural Matings   Avian species generally mate frequently and it is not uncommon for chicken males of the egg type breeds to mate more than 25 times per day. The actual mating is referred to as treading in avian species and after three or four ejaculations, the concentration of spermatozoa becomes quite low. One male can maintain optimum fertility in about 15 hens in the light or egg type breeds of chickens and about 8 to 10 hens in the heavy meat-type breeds.

Treading Process   Many questions are asked by people interested in birds as to the actual treading. During the treading process, the male grasps the female by the loose skin on the back of the neck with his beak to help maintain his balance as he mounts the hen. The receptive female under the weight of the male is in a crouched position. This stance and the weight of the male applies pressure to the abdomen and as pressure is applied on the left oviduct, it pushes across the opening of the cloaca. As the male ejaculates the penis deposits semen on the opening of the oviduct entrance. In gallinaceous birds, the penis does not penetrate the oviduct. Treading takes only seconds and as the male dismounts the female, she rises to a normal standing position and the spermatozoa start moving up the oviduct. Spermatozoa introduced in the vagina of turkey hens may reach the infundibulum within 15 min. The spermatozoa are stored in glands at the junction of the uterus and vagina called uterovaginal sperm storage glands.

Collecting Semen   Most work with semen collection has been done using chickens or turkeys. Semen donor males must be separated from the females and trained for semen collection. The procedure is to grasp the mature male securely by the feet while massaging the area of the back over the testes. The male is then grasped above the cloaca applying a slight pressure to the internal lymph fold and the erected phallus as the tail is pushed forward with the palm of the hand. After several attempts over different days, the males should start ejaculating for semen collection. Males can be used for collection purposes two or three times a week. The semen from several chickens or turkeys is often pooled before insemination.

Semen Storage   The freezing and long-term storage of chicken and turkey semen have not been perfected at this time. Semen can be frozen and thawed immediately but detrimental effects occur; long-term storage doesn't make it any worse. The high concentration of glutamate in avian semen may be one of the problem areas. Currently, semen is collected and utilized within an hour or less of collection time. Diluents or semen extenders are commercially available to use with turkey semen.

Light Effects   The growth and development of the testes generally require 12–14 hr of light. In precocious birds such as the Japanese quail, completion of all stages of spermatogenesis may be complete in as little as 25 days. Green and blue lights have less of a stimulatory effect on the pituitary than red or orange lights. Light intensity is not a major factor within wide limits of 2–50 lux, but dimmer light may retard sexual development.

Male Sex Hormones   The androgens are the primary sex hormones in birds, and as in mammals, testosterone is the most important of these hormones. The androgens are responsible for stimulating sexual desire, growth of accessory reproductive organs, and secondary sex characteristics such as comb growth and color.

Caponization   Caponization is the term used to describe castration in avian species. Capons, which are castrated males, grow more slowly and deposit more fat than uncastrated males. This practice has been used since Roman times to produce superior birds for the consumer. Modern-day production practices have eliminated the need for caponization. Capons can still be found during the holiday season but the number sold is relatively small.

Caponization requires the opening of the body cavity and the subsequent tearing of the air sacs to remove the testes. Young males can best be castrated after 12–24 hr of fasting when they are 3–6 weeks of age.

Sexing   Many methods of sexing newly hatched chickens, turkey poults, and waterfowl have been attempted over the years. There is a very definite need to be able accurately to sex these offspring quickly and without harm.

The most popular method of sexing chicks, turkey poults, and waterfowl is by cloaca eversion. Vent sexing is accomplished by examining the phallic region. Experienced sexers can determine the sex of 1000 chicks per hour with 98% accuracy. Commercial egg-type males are discarded after sexing and often utilized in pet foods. Turkey poults are commercially grown with the sexes raised separately. Young waterfowl are probably easiest to sex at hatching because of the developed penis.

Genetically, crosses can be made that are able to be color sexed. The problem with this method is that a few chicks are difficult to sex correctly, lowering the accuracy.

Another possible method is rapid versus slow feathering. With the use of carefully selected parent lines, the female chicks are rapid feathering, noticeable on the wing primary feathers, and the males are slow feathering (Fig. 3.10).

FIG. 3.10.   Photograph of the primary feathers on the wing of a fast feathering chick.

REFERENCES

Freeman, B. M. 1983. Physiology and Biochemistry of the Domestic Fowl. Academic Press, New York.
Sturkie, P. D. 1976. Avian Physiology, 3rd Edition. Springer-Verlag, New York.
USDA. 1969 (July). Consumer and Marketing Service Egg Grading Manual. Agricultural Handbook No. 75. Washington, DC.
Yapp, W. B. 1970. The Life and Organization of Birds—Contemporary Biology. American Elsevier, New York.

Genetics and Poultry Breeding

The science of genetics deals with the transmission of individual characteristics of both parents to the offspring through the mechanism of heredity. Poultry breeding utilizes genetic principles to accomplish the goals of poultry producers for meat or egg production.

THE CELL THEORY

All plants and animals are made of small building blocks called cells. An individual cell contains a cell wall (membrane) and a nucleus within the membrane. The cellular material, cytoplasm, is found between the nucleus and membrane. The chromosomes, contained within the nucleus, carry genetic material called genes (units of inheritance). In body cells, genes occur in pairs on chromosomes. Chromosomes also occur in like pairs known as homologous chromosomes. The number of pairs of homologous chromosomes is constant within normal individuals in a species. In chickens, it is 39 and the number of chromosomes varies for different species.

In all animals, including birds, the body cells contain one pair of sex chromosomes since they relate to the animal's sex. They are referred to as the X and Y chromosomes with the X being larger than the Y. In humans and farm animals, the male is XY (heterogametic) and the female is XX (homogametic). In birds, the males are ZZ (homogametic) and the females ZW (heterogametic).

	Mammal				Avian	
	X ♀ X				Z ♀ W	
♂ X	XX	XX		♂ Z	ZZ	ZW
Y	XY	XY		Z	ZZ	ZW

All the remaining chromosomes other than the sex chromosomes are known as autosomes. Each individual animal receives one chromosome from its father and one from its mother, constituting a pair of chromosomes. As that animal becomes sexually mature and reproduces, it transmits one of each pair of its chromosomes, determined by chance, to its individual offspring. The diploid ($2n$) number of chromosomes are in body cells. Gametes or sex cells contain the haploid ($1n$) number of chromosomes. Table 4.1 indicates examples of chromosome variation in various animals.

CELL DIVISION

Cells increase in number by dividing, either by mitosis or meiosis. In mitosis each cell divides and forms two cells, each of which possesses the diploid number of chromosomes, the same as the original mother cells. Body cells (autosomes) divide in this fashion.

Meiosis is a type of cell division that occurs in sex cells, in which the chromosome number is reduced from diploid to haploid. This reduction

TABLE 4.1. Chromosome Numbers in Selected Animals

Animal	Diploid	Haploid
Cat	38	19
Cattle	60	30
Chicken	78	39
Dog	78	39
Donkey	62	31
Duck	80	40
Goose	90	45
Horse	64	32
Human	46	23
Turkey	80	40
Swan	84	42
Swine	38	19

ensures that each parent contributes one-half of the chromosomes of the progeny. When fertilization takes place with the union of sperm and egg, the diploid chromosome number is restored in the sex chromosomes the same as in the autosomes (Fig. 4.1).

GENE FUNCTION

The gene is the smallest unit of inheritance and is a portion of a DNA (deoxyribonucleic acid) molecule. This DNA molecule is the foundation of a chromosome. The molecule resembles a lengthy twisted ladder in which the strands (polymers) are joined by rungs.

The polymers are composed of repeated units called nucleotides. Nucleotides are composed of a nitrogenous base (a purine or a pyrimidine) linked to a sugar and then linked to a phosphoric acid molecule (Fig. 4.2).

Ribose is the sugar in RNA, and deoxyribose is the sugar in DNA as shown in Fig. 4.3. Two single strands (nucleotides) of a double-stranded molecule of DNA are joined by hydrogen bonds between bases: adenine always joins thymine and guanine always joins cytosine (Fig. 4.4).

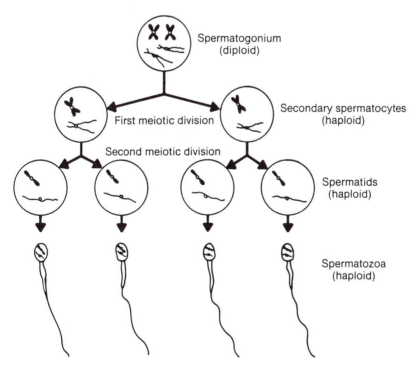

FIG. 4.1. Meiosis—cell division in sex cells.

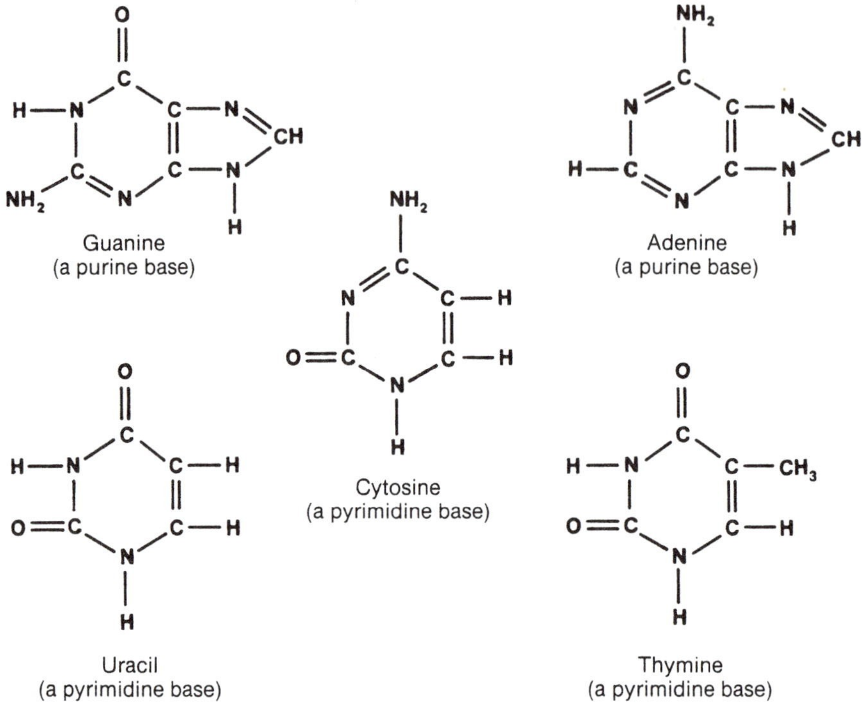

FIG. 4.2. Nucleotides.

A gene is thought to be comprised of a portion of a double-stranded DNA molecule of about 500 or more consecutive base pairs. The gene's function is to replicate itself while a cell is being formed and to send a code to the cytoplasm to build specific proteins. During replication of the polymers of DNA, they separate and each polymer serves as a mold for making the missing part. Base pairing provides a mechanism by which the DNA molecule accurately copies the double-stranded DNA regardless of the duplication number. DNA is found in the cell nucleus. The proteins, however, are synthesized in the cytoplasm by ribosomes linking the amino acids as instructed by the gene. Messenger RNA (mRNA) is made by the DNA to carry instructions since it cannot leave the chromosomes. mRNA travels to the cytoplasm with a message indicating the type of protein to be built.

GENETIC CODE

The genetic information is stored within the DNA molecule in the form of a triplet code called a codon. The code for the formation of a

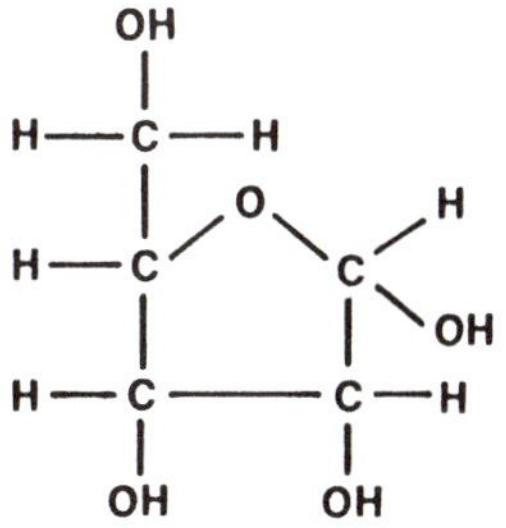

OH
|
H—C—H
| O   H
H—C   C
| OH
H—C ——— C—H
| |
OH   OH

Ribose

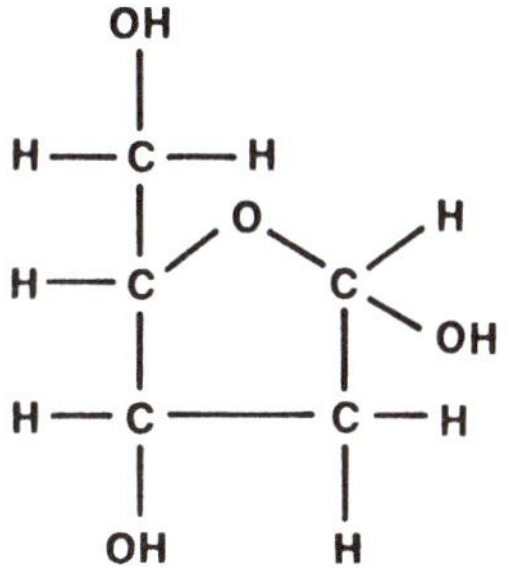

OH
|
H—C—H
| O   H
H—C   C
| OH
H—C ——— C—H
| |
OH   H

Deoxyribose

FIG. 4.3.   Sugar molecules found in RNA (top) and DNA (bottom).

single polypeptide chain of a protein is thought to be carried on a gene as well as instructions to start and stop the amino acid sequence. A long strand of mRNA is formed from the code of one of the strands of the DNA molecule of the gene and is carried to the cytoplasm. mRNA forms the mold for a sequence of amino acids in the ribosomes. Transfer RNA (tRNA) completes the message. A tRNA molecule specifies only one amino acid and recognizes a specific enzyme to join with it. tRNA contains three bases called anticodons that seek and complement a codon on the RNA in ensuring the order of amino acids specified by the DNA (Fig. 4.5).

One theory is that there are structural and control genes. The structural gene functions to make specific protein and enzymes, while the control genes regulate the activity of the structural genes. Control genes fall into two categories: operator and regulator genes. Regulator genes probably produce substances that combine with proteins of the structural genes to produce repressors that keeps the operator gene turned off, inhibiting enzyme formation by the structural gene. The repressor may be blocked by a metabolite (inducer) and allows the operator gene to turn on the structural gene

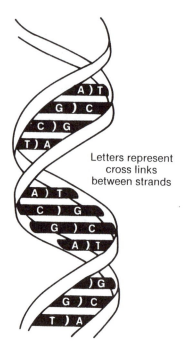

FIG. 4.4.   Nucleotides of a double-stranded molecule of DNA joined by hydrogen bonds between bases.

and produce gene products. The entire gene complex is called the operon.

MUTATIONS

Genes duplicate themselves exactly, generation after generation. However, on occasion mistakes are made in the duplicating process and a new gene (allele) is formed. Most, if not all, mutations are due to a change in the code delivered by mRNA to the ribosomes as sent by the gene. If an altered message is delivered the ribosomes synthesize a different protein, which may cause a defect or new genetic trait to appear. Genetic traits of this type are the raw material with which a poultry geneticist works.

The law of segregation and recombination of genes shows that when two genes are paired in body cells, they segregate independently as demonstrated in the example using Black Rose Comb and White Rose Comb (Fig. 4.6).

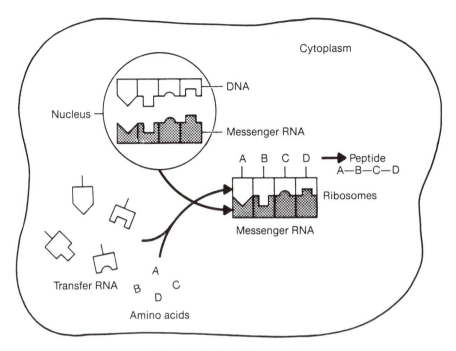

FIG. 4.5.   DNA - RNA sequence.

In the F_1 (first filial) generation, the genes B and b are paired. In the F_2 generation, they segregate in the gametes independently and recombine in pairs randomly.

PHENOTYPIC EXPRESSION (NONADDITIVE)

Phenotype refers to differences we can determine with our senses such as chickens' comb type, color, or clean or feathered shanks, just to name a few. Genotype refers to the genetic makeup of an animal's individual pair of genes such as color type; $BB, Bb,$ and bb all represent potential genotypes. A dominant gene covers or masks the expression of its allele. For example, a dominant white exists in poultry that has been useful in breeding broilers that have white plumage. This makes the pinfeathers white on the carcasses so there is no pigment in the follicle, creating a more attractive carcass. This occurs as follows:

	B	B	
WI	IB	1B	white 0 and 0
I	IB	1B	

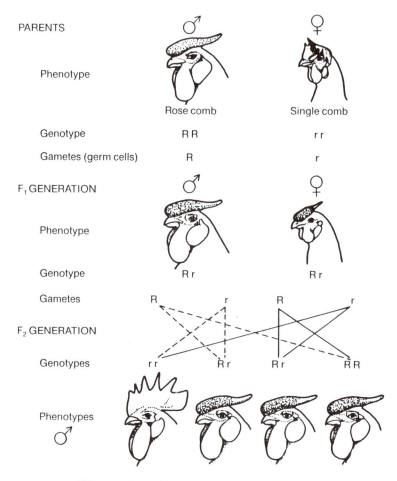

FIG. 4.6.   Law of segregation and recombination.

The recessive traits are always masked or covered when they appear in an *IB* heterozygous (unlike) state. Recessive traits express themselves only in a homozygous (like) state. Examples of other types of gene expression, though not discussed, are partial dominance, overdominance, and epistasis.

PHENOTYPIC EXPRESSION OF GENES (ADDITIVE)

Additive gene action means that the effect of each gene contributing to the phenotype of an animal for a particular trait adds to the phenotypic effect of another gene contributing to the same phenotype. Both additive and nonadditive traits affect economic traits making the geneticist's job more difficult. These economic traits in poultry include rate of gain, feed efficiency, and egg laying. Sex-linked inheritance refers to inheritance carried on the nonhomologous portion of the Z chromosomes. These traits are often recessive and have interactions with sex hormones such as plumage type in male and female chickens. Sex-limited traits are expressed in only one sex but are affected by many genes. Egg laying in chickens is a good example: roosters do not lay eggs but transmit genes influencing egg laying.

Many gene pairs affect expression of traits in poultry having economic values. Fertility, rate of growth, feed efficiency, and carcass quality are but a few. These kinds of traits determine profit or loss in the poultry industry. Many genes are involved and there are no major phenotypic differences from one chicken to another. Inheritance of this type is quantitative.

Qualitative traits refer to things such as feather color, comb type, and similar traits in which one or relatively few pairs of genes are involved. Qualitative traits deal specifically with the individual, whereas quantitative traits involve the total population more and the individual less.

Phenotypic variation is a quantitative trait caused by heredity, environment, and the interaction of the two.

HERITABILITY ESTIMATES

Heritability estimates measure the proportion of the phenotypic variation attributable to heredity expressed in percentages. The heritability estimate when subtracted from 100% gives an estimate of the total phenotypic variation associated with environment (Table 4.2).

The lower the heritability estimate for a trait, the slower the expected progress. Phenotypic variation caused by environment increases as the heritability estimate decreases.

TABLE 4.2. Heritability Estimates of Traits in Chickens

Trait	Percentage
High	
Body weight	60
Egg size	50
Medium	
Growth rate	35
Low	
Sexual maturity	30
Intensity of lay	10
Hatchability	12

SELECTION

Selection is simply permitting selected animals to produce the next generation. Geneticists practice artificial selection as opposed to nature's survival of the fittest or natural selection.

CURRENT BREEDING SYSTEMS

Inbreeding is the mating of closely related individuals over a period of time in a closed flock situation. This can include sister × brother crosses (sib crosses) or mother × son, father × daughter, or cousins. Whatever method of inbreeding is used it increases the homozygosity of that population. The uniformity of genes carried on the chromosomes of inbred birds is higher than in stocks not inbred. Inbreeding, in addition to reducing performance, increases the number of undesirable genes as well as the desirable ones. Inbreeding does offer the opportunity to cross these inbred birds with other unrelated birds resulting in heterozygosity with the crossed birds often showing hybrid vigor.

STRAIN CROSSES

Some well-known breeds of chickens have populations or strains within them that have been developed through closed flock breeding. When crossed with other strains, they perform better than the parent strain. This is referred to as "general combining ability." Some strains cross well only with other particular strains and have specific combining ability. Strain crosses can become very involved: strain A as shown in the diagram works well only as a male line, not as a female line. Many of our modern commercial egg-laying stocks are generated as strain crosses.

$$A \qquad B$$
$$0 \qquad 0$$

F_1 Desirable
egg producer

A popular method of strain crossing is "reciprocal recurrent selection." The aim is to improve the specific combining ability of the strain with selection within the strain dependent on the performance of the strain cross.

CROSSBREEDING

Breeds are groups of animals that are related and tend to be more homozygous for certain traits. In a breed of chickens, for example, there are individuals, though not closely inbred, that are more closely related in a number of characteristics to each other than to individuals in another breed. The main result of crossbreeding is opposite to that of inbreeding. Crossbreeding results in increased heterozygosity and tends to cover up recessive genes, decreases breeding purity, and eliminates families in one generation. The main phenotypic effect of crossbreeding is to cause an improvement in traits related to physical fitness. The increased vigor of crossbred poultry is known as heterosis or hybrid vigor. Hybrid vigor may be expressed as follows:

$$\text{hybrid vigor } (\%) = \frac{F_1 - P_1}{P_1} \times 100$$

or

$$\text{hybrid vigor } (\%) = \frac{\text{crossbred average} - \text{purebred average}}{\text{purebred average}} \times 100$$

Nonadditive gene action, such as dominance, overdominance, and epistasis, is responsible for hybrid vigor. It is not from additive gene action. Currently, crossbreeding is used to produce most of our modern day broiler genetic stocks.

SELECTING SUPERIOR GENETIC STOCK

Regardless of the breeding system used, individuals must be selected for breeding stock. Attention should be given to the performance of the individual when possible. When traits are medium to highly heritable an individual animal's own worth is most important. Since these traits are determined primarily by additive gene action, birds superior in group testing are likely to be superior because of genetic makeup. Low heritability traits are greatly affected by environment and nonadditive gene action. Individuals recognized for superior traits often are disappointing in performance as parents because of the variation caused by environment.

When many individuals are compared under similar environmental conditions such as in a broiler house, superior performance likely indicates a superior genetic makeup. An example in broilers is rate of growth, which is moderate to highly heritable and affected by additive genes. Mass selection could be practiced for this single trait by using only the males in the top 20% of the body weight range.

PROGENY TESTING

Progeny testing is a tool used in poultry breeding both for egg and meat production stocks. Progeny testing for traits such as egg laying is a long, drawn-out procedure that requires extensive record keeping. When selection proceeds to the point at which multiple traits are selected for simultaneously, the rate of progress is greatly reduced. Many breeding companies today utilize computers to analyze the mass of information gathered from the various pens of flock matings retained for breeding purposes.

SOURCES OF GENETIC STOCK

Supplying commercial poultry breeding stock is a big business today. The large breeding companies that have evolved over the last 30 years have in most instances come from successful small poultry breeders. These small breeders had a demand for the superior genetic stocks they had developed. Over the years, they added the scientific personnel necessary to do the research to maintain a superior genetic stock. Within these companies strong marketing organizations were developed to generate greater sales. Smaller companies went out of business or were absorbed by larger companies. Today essentially 10 companies supply a majority of the world's poultry genetic stock. It appears that one of these companies may sell 20% of the world's commercial egg layers. These large poultry breeding companies are now frequently part of larger conglomerate business ventures often dealing with pharmaceuticals or similar products.

BREEDS AND VARIETIES OF CHICKENS

In 1874, the first Standard of Perfection was issued by the then newly formed American Poultry Association to establish a standard of excellence for exhibition of poultry. The Standard of Perfection poultry was divided into several categories. Class was a category that indicated the origin or area of development such as the American, Mediterranean, or Asiatic class. Breed was generally accepted as referring to body shape and size. An example would be the Cornish breed from the English class. Variety was a subcategory of breed and generally referred to color pattern and comb type. The Single Comb White Leghorn is an example of a variety of the leghorn breed from the Mediterranean class. Interest in exhibition poultry peaked around the turn of the century with over 200 breeds and varieties recognized in the various classes. This early interest in attempting to get true breeding breeds and varieties through phenotypic selection helped to expand the genetic pools available to geneticists in later years.

THE UTILITARIAN CONCEPT

The American class of chickens that evolved had as its genetic base stocks from the English, Mediterranean, and Asiatic classes. Birds in these breeds and varieties of chickens were intermediate in size and possessed good egg laying and carcass qualities. Barred Plymouth Rocks, Rhode Island Reds, and White Plymouth Rocks were popular breeds of the type known among poultrymen as dual-purpose breeds. These breeds worked well in small flock situations in this country during the 1930s and 1940s. As poultry farms increased in size in the early 1950s, greater efficiency was needed to produce poultry meat and eggs at a profit.

In the early 1950s, the "utilitarian concept" or use concept was employed by poultry geneticists. Genetic selection diverged from the American dual-purpose bird concept. A chicken would be either meat type or egg type. Egg-type stocks that evolved over the last 30 years have primarily been of the leghorn type. Through the use of inbreeding and various methods of strain crossing, a persistent high rate of lay has been attained coupled with good feed efficiency and large egg size from relatively small-bodied hens.

The genetic package consists of a female line from the American class that is essentially of White Plymouth Rock ancestry. This female type was selected because of desirable egg-laying ability and plumage color coupled with a fairly meaty carcass. The male line is primarily composed of Cornish bloodlines from the English class because of the high meat yield and superior body conformation (Fig. 4.7).

Several items should be explained concerning the cross used in Fig. 4.7. In the parent stock the White Plymouth Rock males are not used as breeders in the reciprocal cross with the Cornish females. Cornish females are not good layers and the cost per chick produced would not be competitive with the other cross. The F_1 offspring or commercial meat-type chicks will exhibit hybrid vigor, be heterozygous for numerous traits, and outgrow their parents. However, since they are heterozygous, they will not breed true and are considered a terminal cross. If similar birds are desired one must go back to the foundation parent stock and repeat the cross. A similar situation exists in commercial layers since the commercial layer is the end product of strain crosses. This allows commercial breeding companies to retain the superiority of their grandparent and parent stocks without fear of their being reproduced through the offspring. Commercial meat- and egg-type chickens can then be supplied through dealers and hatcheries franchised by the primary breeding company throughout the world. Poultry breeders through heterosis have a type of obsolescence built into the commercially available chicks.

FUTURE POULTRY GENETICS AND BREEDING

The rate of genetic progress has declined in recent years as the physiological limits of the birds for some traits are approached. Many recent production gains have been achieved by more closely controlling the chicken's environment. Additional gains will be achieved with further knowledge of the interaction of genetics and environment. Poultry genetics is becoming increasingly more complex, and no one geneticist today can expect to be competent in all areas. The genetics

FIG. 4.7. Commercial broiler-type cross.

research of the future will encompass genetics, biochemistry, immunology, and greater use of computer technology, as well as the genetic engineering area known as biotechnology.

LETHAL GENES

Lethal genes exist in the gene pool only when homozygous or hemizygous conditions occur, such as in a sex-linked lethal. Lethal from a genetic standpoint implies that the deviation from normal is so extreme that the individual bird cannot survive. Some lethal genes in birds are not recognized because no easily visible abnormalities occur or they are lethal during embryo development and hatching does not occur. Table 4.3 lists some of the lethal genes found in poultry. Lethal genes are generally undesirable and should be eliminated from the breeding population. The method used to identify these genes depends on the type of gene involved. For example, sex-linked lethals can be carried only by males. All male carriers and progeny should be eliminated as soon as detected. Most lethals are autosomal recessives and can be detected only by backcrosses to sires or dams.

PARTHENOGENESIS

Parthenogenesis is the initiation of cell division in an unfertilized egg without genetic contribution from a male gamete. In most instances, parthenogenesis ends in an abortive type of early development but may continue to include formed embryos and, in rare instances, in chickens and turkeys, hatched parthenogens. Parthenogenesis is a naturally occurring phenomenon in many animal groups. Among invertebrates, particularly insects, parthenogenesis is an efficient method of reproduction.

TABLE 4.3. Examples of Lethal Genes on Poultry

Gene	Characteristic
Creeper	Shortened tibia or metatarsus—inability to emerge from the shell
Crooked-neck dwarf	Dwarfism-skeletal abnormalities
Curled beak	Inability to pip
Sex-linked lethal	Early embryo death
Sex-linked recessive lethal	Cartilaginous skeletal abnormalities in females
Polydactyly	Abnormal number of toes—inability to emerge from the shell
Open-breast syndrome	Absence of sternal crest—breast muscle latrophy

Source: Landauer (1973)

When parthenogenesis occurs in higher vertebrates, cell divisions are initiated but mitosis fails to continue for very long and most embryos die. Some parthenogenic turkey poults have been hatched by Olsen *et al.* at the USDA laboratory at Beltsville, Maryland. These poults and other potential avian parthenogens have several potential uses in embryological studies, lethal genes, mutations, and skin grafts. Parthenogens could also be used for the development of isogenetic lines since they may be homozygous at all or most loci.

REFERENCES

Campbell, J. R., and Lasley, J. F. 1969. The Science of Animals That Serve Mankind. McGraw-Hill, New York.

Hutt, F. B. 1949. Genetics of the Fowl. McGraw-Hill, New York.

Hutt, F. B., and Rasmussen, B. A. 1982. Animal Genetics, 2nd Edition. John Wiley, New York.

Jull, M. A. 1951. Poultry Husbandry. McGraw-Hill, New York.

Landauer, W. 1973. The Hatchability of Chicken Eggs as Influenced by Environment and Heredity. Storrs Agricultural Experiment Station, The University of Connecticut, Storrs, CT.

Moreng, R. E., and Avens, J. S. 1985. Poultry Science and Production. Reston and Prentice-Hall, New York.

5

Incubation and Hatchery Management

POULTRY REPRODUCTION

Poultry is suited to mass production techniques because poultry numbers can be expanded faster than other livestock species by artificial incubation. Once an egg is laid, all ties with the mother are severed. From that point on, the embryo develops independently, since a complete balanced set of nutrients is stored in each egg.

The ovum on the yolk of an egg is fertilized no later than 15 min after ovulation. Fertilization occurs in the infundibulum near the junction of the magnum. Although several sperm can penetrate the ova, only one can fertilize it. About 5 hr after ovulation, as the egg enters the isthmus, the first cell cleavage of the developing embryo occurs followed by a second one 20 min later. While the egg is in the isthmus, it may even cleave to the four- or eight-cell stage. About 9 hr after ovulation, the blastoderm (developing embryo) has grown to the 256-cell stage so that by the time the egg is laid, the formation of the gut (gastrulation) is usually complete in the embryo. Normal cellular division continues in the developing embryos after laying as long as the egg temperature remains above approximately 80°F (26.8°C) (physiological zero). When the egg is cooled below physiological zero, cell division stops and the embryo becomes dormant.

STORAGE AND SELECTION OF HATCHING EGGS

Hatching eggs should be gathered three to five times daily to prevent the eggs from incubating, or being broken by the hens, and the hens from becoming broody. It is important that eggs be cooled at least below physiological zero as soon as possible to stop cell division. If the eggs remain warm and cell division continues beyond a certain stage of development and then the eggs are cooled again, embryonic death results.

Hatching eggs are selected and sorted commercially as they are gathered on the breeder farm. Eggs laid on the floor, obviously dirty eggs, as well as cracked and misshapen eggs are eliminated for use as hatching eggs. Uniformity of size is also an important criteria in selecting hatching eggs. Very small eggs and oversized eggs are sold for human food. Generally, the larger the egg the longer the incubation period. Large eggs, compared to other eggs produced in the same flock, will take about 12 hr longer to hatch than smaller ones. Double-yolked and oblong eggs as well as eggs showing signs of poor shell quality, such as thin, porous, or sandy shells, should not be set. These types of eggs have excessive water loss during incubation, although control of humidity during storage and incubation can reduce the problem of moisture loss to some extent.

Most large hatcheries set eggs two to four times a week and use one other day for cleaning the hatchery and incubators. Eggs from an individual flock of hens are normally collected three to five times a day and set twice a week so the eggs will not be held over 4 days. Generally, such eggs are held in an egg room kept at 65°F (18°C) and 75% relative humidity to prevent excessive weight loss.

Storage times of over 1 week cause a decline in hatchability that becomes more pronounced as the length of storage increases. Generally, eggs stored at lower temperatures for prolonged periods deteriorate less in quality than those stored at higher temperatures. An egg room temperature of 60°F (16°C) is ideal for hatching eggs stored for 1 week. Eggs stored for 10–14 days at 50–55°F (10–13°C) hatch better than those stored at higher temperatures. In any case, hatching eggs should not be stored at temperatures under 50°F even though market eggs are often held at temperatures as low as 35°F (2°C) to extend shell life and maintain quality.

All hatching eggs are normally brought to room temperature prior to setting to prevent sweating. The gradual warming also is less of a physiological shock to a developing embryo as it changes from a dormant state to one of active cell division. The hatchability of eggs stored at cold temperatures can be improved by prewarming them for 12–18 hr prior to setting.

Hatching eggs are stored with the large end up. In commercial operations in which storage is generally less than 1 week, the storage position of the egg has little influence on hatchability. However, eggs stored longer than 2 weeks hatch better when they are turned daily. Turning is done to prevent the embryo and the yolk from adhering to the shell membrane which will cause the death of the embryo from dehydration.

Hatching eggs shipped by air do not hatch well when placed directly in incubators after arrival. However, if the eggs are allowed to stand for 24 hr and then slowly warmed to room temperature, hatchability will be similar to eggs that have not been shipped.

In experimental situations, hatching eggs have been stored 3 to 4 weeks in sealed plastic bags flushed with nitrogen or carbon dioxide. Even though such eggs hatched better than controls, storage time beyond 10–14 days does not appear practical for commercial poultry.

EMBRYONIC DEVELOPMENT

The blastoderm is differentiated into two layers of cells by the time the egg is laid. Gastrulation is accomplished by the growth of cells along one portion of the margin of the blastoderm forming a second layer of cells. This inward growth of the second layer of cells eventually divides the blastoderm in two. The blastoderm then divides into the ectoderm above and the entoderm below. After incubation begins, the mesoderm, or third germ layer, differentiates by growing into the blastocole between the other two layers. The ectoderm later forms the lens and retina of the eye, the nervous system, the skin, beak, feathers, and claws as well as the linings of the vent and the mouth. Reproductive and excretory organs, bones, muscles, and blood develop from the mesoderm. The entoderm gives rise to respiratory and secretory organs in addition to the linings of the digestive tract.

As incubation begins, one of the first visible changes on the growing embryo is the development of the primitive streak. The primitive streak comes from two thickenings of cells in the ectoderm close to the starting point of the entoderm. It marks the longitudinal axis of the body of the embryo. Although the embryonic axis is uniform, it is not fixed.

Cells of the area pellucida show uneven growth that gives rise to a series of folds that involves various germ layers. The embryo proper is marked off by these folds. The head fold raises the anterior portion of the embryo above the rest of the blastoderm. The tail fold goes under the posterior extremity of the body to raise it. These folds then join with the lateral folds to mark out the sides of the embryo and eventually lift it above the yolk. Only a narrow stalk connects the yolk and embryo at this time.

EXTRAEMBRYONIC MEMBRANES

The developing embryo has no connection to the mother after oviposition. To compensate for this, four specific membranes develop in the egg enabling the embryo to utilize the nutrients available in the egg and to begin respiration as shown in Fig. 5.1.

Yolk Sac

The yolk sac is formed by a layer of entoderm and mesoderm growing over the surface of the yolk. The uptake of yolk material is accomplished by a glandular absorbing epithelium lining the walls of the yolk sac. The yolk sac or remaining yolk material is pulled into the body cavity just prior to hatching. The unused yolk portion left after hatching is a temporary nutrient source available to the chick to sustain it as it learns to find food for itself. Also, as the lipid materials of the yolk are metabolized, water is produced as a by-product and utilized by the chick. The nutrients left in the yolk normally can

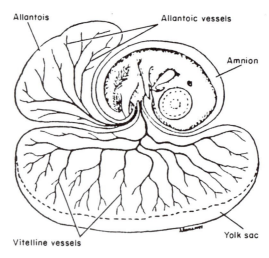

FIG. 5.1. Seven-day-old embryo with its embryonic membranes and embryonic blood vessels.

sustain a chick about 96 hr after hatching and allow newly hatched chicks or poults to be shipped great distances before being placed on feed and water.

Amnion–Chorion

The amnion and chorion both evolve from a fold of the extraembryonic tissue in the head region. This tissue eventually encompasses the whole embryo. It consists of both ectoderm and mesoderm and grows up over the embryo and then fuses at the top. The outer part has ectoderm above and mesoderm below it and forms the chorion which fuses the inner shell membrane with the allantois, aiding the allantois in completing metabolic functions. The amnion is a membranous sac filled with a colorless watery fluid that allows the embryo to float about freely as it develops. It also protects the embryo from trauma.

Allantois

After about 4 days incubation the entoderm from the hind gut forces a layer of mesoderm ahead of it into the extraembryonic membrane. This layer forms the allantois, a highly vascular sac that occupies the space between the amnion and chorion. When the allantois fuses with the chorion, its capillaries come in direct contact with the shell membranes. The allantois has several functions. It is an embryonic respiratory organ that exchanges gases, absorbs albumen to use as a nutrient for the embryo, and also absorbs calcium from the shell for the skeletal formation of the embryo. Another vital function of the allantois is to receive and wall off excretions from the kidney preventing them from contaminating the remaining food source of the embryo.

DAILY EMBRYONIC GROWTH

Day 1
    Development of area pellucida and area opaca on germ spot
    Primitive streak—longitudinal axis of embryo
    Central nervous system as neural groove
    Foregut
    Blood islands in area opaca
    Primordial sex cells appear
Day 2
    Embryo begins to turn on left side
    Heart formation
    Heart beat at 30 hr

Primary division of brain
Forming of eyes
Forming of ear pits
Forming of tail bud
Blood vessels appear in the yolk sac
Day 3
Limb buds formed
Esophagus started
Circulatory system rapidly increased
Day 4
All body organs present, vascular system clearly evident
Day 5
Proventriculus and gizzard formed
Reproductive organs differentiate
Day 6
Differentiation of embryo; becomes bird-like in shape
Main divisions of legs and wings
Body begins to develop more rapidly
Voluntary movement
Duodenal loop formed
Beak and egg tooth take normal form
Day 7
Ceca starts development
Indications of digits
Abdomen more prominent due to development of viscera
Rapid body movement evident
Day 8
Feather genus appears in tracts
Ceca start
Eyes prominent but midbrain no longer as prominent
Day 9
Contour of body bird-like
Allantois nearly surrounds embryo, amnion, and yolk
Day 10
Forelimb-like wing and digits of feet completely separated
Beak starts to harden
Day 11
Abdominal wall appears
Intestines seen
Day 12
Appearance of some down on embryo
Day 13
Down covers body and color from down seen through the amnion
Scales and nails appear

Day 14

Embryo parallel to long axis—head to large end in normal hatching position

Day 15

Small intestines taken into abdomen

Day 16

Scales and nails horny

Albumen nearly gone and yolk increasingly important as source of nutriment

Day 17

Amniotic fluid decreasing

Embryo begins preparation for hatching, beak under right wing toward the lower part of the air sac

Day 18

Amniotic fluid gone—yolk sac ready to enter body

Day 19

Yolk sac starts to enter body through umbilicus

Beak pierces air sac and lungs and air sacs begin to function

Day 20

Pulmonary respiration starts

Allantoic circulation and respiration cease

Allantois begins to dry up

Eggshell is pipped

Day 21

Normal hatching

Forepart of body toward large end of egg

Head under right wing

Legs up under head

Upper mandible with pointed horny cap—egg tooth

Pipping completed—through change in posture of bird's head

The pipping and emergence from the eggshell are accomplished over a period of 10–20 hr. The actual pipping is accomplished about two-thirds of the way up the shell toward the large end of the egg, which gives the chick the largest emerging circumference. The chick breaks the shell by retracting its head and striking the shell with the egg tooth. Eventually a hole is broken in the shell and then the egg tooth is used to pull in shell segments in a counterclockwise motion until there is a fissure completely around the shell. The chick then extends its legs and pushes itself from the shell, wet and exhausted. After several hours, it dries off and is able to walk. Incubation periods of various birds are shown in Table 5.1. In Fig. 5.2, the daily changes in the weight and form of the developing chick embryo are shown.

TABLE 5.1. Incubation Period and Incubator Operation for Eggs of Domestic Birds[a]

Requirements	Chicken and bantam	Turkey	Duck[b]	Muscovy duck	Goose	Guinea	Pheasant	Peafowl	Bobwhite quail	Coturnix quail	Chukar partridge	Grouse
Incubation period (days)	21	28	28	35–37	28–34	28	23–28	28–30	23–24	17	23–24	25
Still-air operating temperature[c] (°F, dry bulb)	$101\frac{3}{4}$	$101\frac{1}{4}$	$101\frac{1}{2}$	$101\frac{1}{2}$	$101\frac{1}{4}$	$101\frac{1}{4}$	$101\frac{3}{4}$	$101\frac{1}{4}$	$101\frac{3}{4}$	$101\frac{3}{4}$	$101\frac{3}{4}$	$101\frac{3}{4}$
Humidity (°F, wet bulb)	85–87	83–85	84–86	84–86	86–88	83–85	86–88	83–85	84–86	84–86	80–82	82–86
Do not turn eggs after day	19	25	25	31	25	25	21	25	21	15	21	22
Operating temperature during last 3 days of incubation[c] (°F, wet bulb)	101	$100\frac{1}{2}$	$100\frac{3}{4}$	$100\frac{3}{4}$	$100\frac{1}{2}$	101	101	$100\frac{1}{2}$	101	101	101	101
Humidity during last 3 days of incubation (°F, wet bulb)	90–94	90–94	90–94	90–94	90–94[d]	90–94	92–95	90–94	90–94	90–94	90–94	90–94
Open ventilation holes one-fourth on day	10	14	12	15	1	14	12	14	12	8	12	12
Open ventilation holes further if needed to control temperature on day	18	25	25	30	25	24	20	25	20	14	20	21

[a] From *Incubating Eggs of Domestic Birds*, Circular 530, June 1972, Clemson University.
[b] It has been reported that duck eggs hatch better in still air incubators than in forced-air incubators.
[c] For forced-air incubators subtract 2°F for the recommended operating temperatures.
[d] Better hatchability may be obtained if goose eggs are sprinkled with warm water or dipped in lukewarm water for half a minute each day during the last half of the incubation period.

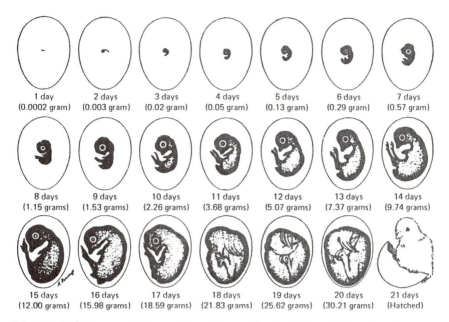

FIG. 5.2. Daily changes in the weight and form of the developing chick embryo. Source: Anonymous (1976).

EMBRYONIC NUTRITION

The nutrients needed for embryo development come from the carbohydrates, fats, and protein in the yolk and albumen of the egg. Carbohydrates probably furnish the energy the first 4 days, but after that, urea is formed indicating protein metabolism. Fat from the yolk is probably the energy source late in the incubation period. Calcium transferred from the shell to the embryo is the most important mineral involved in embryo development. The calcium content in the embryo increases markedly after the twelfth day of incubation as the embryo goes into a rapid growth stage and skeletal development nears completion.

EMBRYONIC COMMUNICATION

In several species of birds, eggs laid in a sequence of several days hatch during a short period of time even though some eggs would have been incubated longer than others. Hatching is synchronized through embryonic communication. The communication is accomplished through vibrations or clicks made by embryo movement. Hatching can be accelerated or retarded by using artificial clicks. Communication of this type has been found in various types of quail. The vocal stimuli of wood duck hens can synchronize the sequence of hatching by retarding some embryos and accelerating others. Perhaps someday, synchronization of hatching can be accomplished in large-scale hatcheries.

EMBRYOLOGY IN RESEARCH AND TEACHING

The development of the chick embryo continues to be a popular method for teaching embryology. Chick embryo studies have been developed for various age levels starting with youngsters in 4-H clubs, numerous science fair projects, and laboratory exercises in college. The chick embryo can be produced in large numbers at various stages of development at relatively small expense.

The chick embryo is also widely used in research, the production of vaccines, and the diagnosis of some avian diseases. The developing embryo is an ideal media for virus production and can be used to prepare specific antisera. Experimental applications include nutritional and cellular differentiation as well as the development of organ systems. Currently, embryos are being used in large numbers to determine the mutagenic or teratogenic effect of numerous chemical compounds being developed for commercial use.

Each year newer and more sophisticated techniques for embryo culture are developed which greatly expand the role of the chick embryo in research. Recent advances in biotechnology and genetic engineering point toward an increasing interest in embryo use in research.

ARTIFICIAL INCUBATION

The foundation of the modern poultry industry is artificial incubation in which mechanical equipment is used to replace the broody hen for egg incubation. The art of incubation has been known for several thousand years, but it has been employed on a commercial scale only within the last 60–70 years. The Chinese developed the first incubators by using a mud barrel heated with charcoal; the eggs were turned by hand. The Egyptians were the first artisans to construct large incubators constructed of brick and heated by fires built in the same rooms in which the eggs were incubated. The incubators held as many as 90,000 eggs and were operated on a toll basis—two chicks returned for each three eggs set (Fig. 5.3). The success of these in-

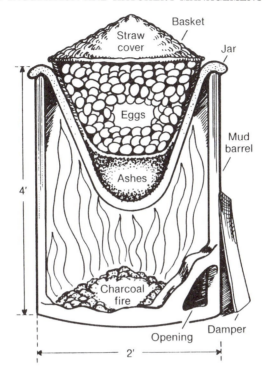

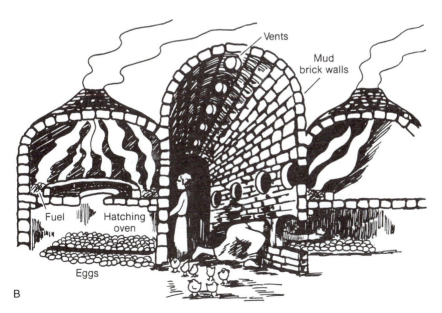

FIG. 5.3.   (A) Sketches of ancient Chinese and (B) Egyptian incubators.

cubators was caused partially by their geographic location. Since they were located close to the equator, the ambient temperature approached normal incubation temperature so little supplemental heat was required. The high humidity in the area was also ideal for incubation. The operators moved and turned the eggs frequently to maintain good incubation conditions. Some of these hatcheries still exist today.

MODERN INCUBATORS

The first large incubator in the United States was built in 1895 by Charles A. Cyphers. It was a 20,000 egg capacity, room-type incubator for duck eggs. S. B. Smith patented the first forced-air incubator in 1918. The forced-air system is still used in all modern day commercial incubators. In 1923, the Petersime Incubator Company introduced the first all-electric incubator. Incubators available today have utilized the latest in modern technology and have greatly reduced the amount of labor required to set and transfer hatching eggs.

Modern day hatcheries are impressive. By using banks of incubators, as many as one million chicks a week can be hatched. A modern hatchery is an example of an engineering solution used to solve a biological problem. The biological requirements of temperature, humidity, air supply, and egg turning have been determined and are built into the incubation system which operates with automated controls.

TEMPERATURE

Developing embryos are quite sensitive to their surrounding environmental temperature during incubation. Temperature regulation is probably the most critical factor in incubation. The optimum incubation temperature appears to be 98.6–100.4°F (37.3–38.3°C) in forced-air incubators and approximately 2°F (1.1°C) higher in still-air incubators. Eggs will usually hatch if kept at temperatures between 95 and 104°F (36–40°C). Embryos cannot tolerate elevated temperatures. Even short periods of slightly increased temperature increase mortality. Continuously high incubation temperatures increase the number of abnormal chicks as well as mortality. If incubation temperatures reach 115°F (46°C) for 3 hr, or 120°F (49°C) for 1 hr, all embryos will die. Chicks subjected to a severe heat stress have an unsteady gait and wiry down. Since embryos become exothermic as they develop, incubators must be able to cool eggs as well as heat them. Temperature effects on embryos are shown in Fig. 5.4. Alarm systems are installed in all large modern incubators to signal increases in temperature above the normal operating range. In addition, all modern hatcheries have standby generators capable of running the hatchery if a power failure occurs.

Low incubation temperatures can also affect development. Low temperatures slow cell division and greatly increase the incidence of abnormal cell divisions. Short periods of cooling are not thought to adversely affect hatching since this occurs in natural incubation as the hen feeds, but embryos are particularly susceptible to cooling during the last 2 days of incubation. For these reasons, thermometers should be checked for accuracy routinely against a test thermometer. Thermometers with separated mercury do not read accurately and should have the mercury reunited before use. Embryos developing in an incubator have a high heat requirement during the first 13 days of incubation; they then generate heat that needs to be dissipated.

RELATIVE HUMIDITY

Comparing the temperatures recorded by wet-bulb and dry-bulb thermometers supplies data to calculate relative humidity. As air temperature rises, the capacity of air to absorb and hold moisture increases. The amount of moisture in the air around the egg must be controlled to a normal evaporation rate to prevent too rapid a moisture loss or retention of too much water in the egg. The size of the chick at hatching is affected by moisture loss. If eggs dry out quickly, the chicks

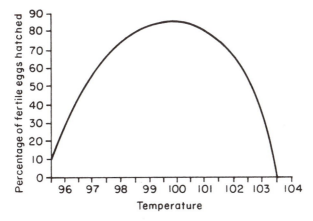

FIG. 5.4. The effect of incubation temperature on percentage of fertile eggs hatched. Relative humidity 60%, oxygen 21%, carbon dioxide below .5%. Source: Anonymous (1976).

TABLE 5.2. Relative Humidity

Incubator temperature (°F)	Wet-bulb readings					
100	81.3	83.3	85.3	87.3	89.0	90.7
101	82.2	84.2	86.2	88.2	90.0	91.7
102	83.0	85.0	87.0	89.0	91.0	92.7
Relative humidity (%)	45	50	55	60	65	70

Source: Anonymous (1976).

are small; when humidity is high, abnormally large mushy chicks are produced.

The optimum relative humidity for incubation during the first 19 days is between 50 and 60% (Table 5.2). Normal hatching eggs should lose about 10.5% of their weight before transfer at 19 days. The humidity in the hatcher should be gradually increased to a relative humidity of 75% at the peak of hatching. As the relative humidity is changed upward, the temperature is decreased slightly.

Egg size and shell quality also affect weight loss in hatching eggs. Large eggs have less shell surface area per unit of weight than small eggs. Porous, thin-shelled eggs lose more moisture than thicker-shelled ones. The air cell should increase in size as incubation progresses as shown in Fig. 5.5.

AIR SUPPLY

As the embryo develops, it requires an adequate supply of oxygen and a mechanism to exhaust carbon dioxide and water through the shell membranes and pores. Air contains about 21% oxygen at sea level so that in forced-air incubators, the oxygen level is seldom a problem. When good hatches occur in large hatchers, large amounts of carbon dioxide may be produced. Hatchability can drop about 5% for each 1% drop in oxygen. As embryos reach a more advanced stage of development, the oxygen requirement increases and more carbon dioxide is produced.

Carbon dioxide should be maintained at levels naturally occurring in the air, which is about .05%. Levels higher than this start to reduce hatchability. Two percent carbon dioxide in the air greatly reduces hatchability. Since carbon dioxide is a natural by-product of embryo metabolism, all incubators should have recorders to monitor CO_2 levels.

It is well documented that hatchability decreases as the altitude increases. The decrease is seldom noticed at levels below 2500 ft;

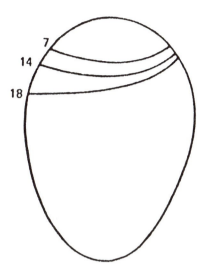

FIG. 5.5.   Diagram showing the air cell on the seventh, fourteenth, and eighteenth days of incubation. Source: Anonymous (1976).

however, when the altitude is above 3500 ft, the loss is considerable. Embryo growth is retarded by low oxygen levels but the probable cause of death comes from reduced hemoglobin values. Oxygen injected directly into the incubator can be used in an attempt to correct the effects of high altitudes on hatchability.

HATCHING EGG POSITIONS

Under natural conditions, a hen turns eggs frequently, using her body as she settles on the nest and then her beak as she turns eggs under her body. The shape of the egg and the dished shape of the nest permit the egg to lay in a horizontal position with the large end slightly elevated.

This egg position permits the embryo to develop in the large end of the egg with the head uppermost toward the air cell which then permits the chick to pip satisfactorily and emerge from the shell at the largest opening (Fig. 5.6). Successful artificial incubation is accomplished by recreating these situations by mechanical means in the incubator. For example, eggs should be incubated with the large end up to prevent the embryo from developing in the small end of the egg away from the air cell and where a smaller shell opening would cause difficulty in emerging. Hatchability is decreased about 10% by setting

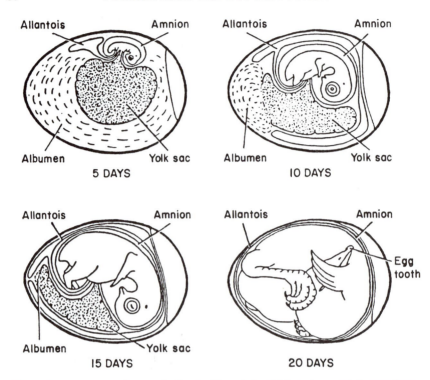

FIG. 5.6. Successive changes in the position of the chick embryo and its membranes.
Source: Anonymous (1976).

eggs small end up and chick quality declines considerably. If eggs are
not turned in the incubator, the yolk rises through the albumen layers
as the specific gravity of the albumen lessens and comes in contact
with the inner shell membrane where it may stick to the shell and
eventually cause the death of the embryo. In most commercial incubators, the eggs are set large end up and are rotated back and forth
45° from vertical then reversed 45° to the opposite side. Rotation of less
than 45° is not adequate for optimum hatchability. There does not
appear to be any advantage in turning eggs more frequently than six
to eight times daily, but the turning process should be completed
quickly whenever it is done.

   The chick's urge to pip and emerge from the shell is likely initiated
by a decrease in oxygen supply and an increase in carbon dioxide
within the egg as pulmonary respiration begins. Oxygen content of the
air cell is only 15–16% while the carbon dioxide level may be as high
as 4%.

EMBRYONIC MORTALITY

Mortality rates vary with each group of eggs incubated. They can be caused by many things. Normal mortality generally has two peaks during the incubation period. These mortality peaks occur during early incubation—days 2, 3, and 4—totaling around 1.5% and late in the incubation period—days 19, 20, and 21—totaling around 3.0% in normal hatches. The early peak is associated with physiological adjustment of the embryo as the various systems of the embryo are initiated. The second peak is associated with pulmonary respiration and the hatching process. Faulty hatcher management can greatly increase embryo mortality during the second period. A guide to incubation troubleshooting is given in Table 5.3.

THE NATIONAL POULTRY IMPROVEMENT PLAN

Disease organisms adversely affect developing embryos, hatchability, and chick quality. Some disease organisms can establish themselves in the hatchery and incubation equipment and then infect other sets of eggs. Still other organisms affect the parent flock. The disease organism can then pass from the parent hens through the egg and affect the chicks. Disease transmission was a severe problem during the time the commercial hatchery system was developing in the United States. For this reason, the U.S. Department of Agriculture,

TABLE 5.3. Incubation Trouble-shooter

Symptom	Probable cause
Chicks hatching too early, with bloody navels	Incubator temperature too high, communication problems among embryos
Draggy hatch: some chicks early but slow in finishing	Temperature too high
Delayed hatch: eggs not pipping until twenty-first day or later	Temperature in incubator too low
Short down on chicks	High temperature or low humidity
Mushy chicks; dead on platform; bad odor	Navel infection caused by bacteria in incubator, high humidity
Chicks too small	Low humidity or high temperatures
Shell sticking to chicks	Low humidity at hatching time
Chicks smeared with egg contents	Low average temperature; humidity may be too high
Spraddled legs	Slippery hatching surface in trays
Crippled chicks: missing eye, cross beak, extra leg, etc.	Mostly chance; poor nutrition of parent stock; heredity
Rough navels	High temperature or low humidity

Source: Anonymous (1976).

cooperating with individual states, initiated a program in 1935 known as the National Poultry Improvement Plan. The objective of the plan was to improve the production and market qualities of chickens and reduce losses attributed to hatchery-disseminated diseases. A sanitation plan was formulated to aid hatchery operators and a blood testing program was established to identify breeders carrying disease so they could be eliminated. The initial work in this program aided in identifying a bacterial disease transmitted from the hen to the chick through the egg caused by the bacterium *Salmonella pullorum*. Today, breeders are not only carefully screened for this disease but they are also screened or vaccinated against a host of other diseases including pullorum, Arizona, fowl typhoid, infectious bronchitis, Newcastle, *Mycoplasma gallisepticum,* and *Mycoplasma synoviae.*

HATCHERY SANITATION

Hatchery sanitation aids in breaking the disease cycle. Many compounds are available for use in a hatchery sanitation program. They include cresols, phenols, codeine, chlorine, quaternary ammonium compounds, and formaldehyde. All of these compounds have their good and bad points; however, disinfectants used should be (1) readily available and inexpensive, (2) soluble in water, (3) nonstaining and noncorroding, (4) nontoxic to humans and animals, (5) effective in the presence of organic matter, (6) free from strong odors, (7) able to penetrate materials and cracks, and (8) highly germicidal.

FRANCHISE HATCHERY SYSTEM

Some hatcheries emerged with superior breeding stock coupled with good management and thrived because they sold chicks of outstanding quality. Because these hatcheries grew and prospered, they eventually attracted the interest of large agricultural conglomerates which recognized the business potential of supplying quality poultry breeding stock. They also recognized the potential for national and international sales and accomplished this through a franchise system. Generally, a territory was established and an individual agreed to build a hatchery. This person was then supplied with foundation breeding stock to produce chicks for a particular company. The individual franchise holder was assured quality breeding stock and aided by the parent company in sales promotion and advertising. In return the franchise holder returned a percentage of the gross sales to the parent company to compensate them for their efforts in developing superior lines and to generate profits and money to do research to develop future generations of foundation stock.

HATCHERY SERVICES

In addition to the incubation and hatching of eggs, modern hatcheries also can provide a number of hatchery-associated services. Commercial egg-type chicks are sexed at the hatchery, generally by cloacal eversion, and only the pullets are saved because the male egg-type chicks do not grow efficiently enough to be used for meat purposes. Broiler chicks and turkey poults may also be sexed at the hatchery for sex-separate rearing. Most chicks, regardless of type, are debeaked at the hatchery to prevent cannibalistic tendencies that may occur later. Most chicks are vaccinated against Marek's disease at the hatchery. Other services may include detoeing, dubbing (comb removal), or inoculating against specific diseases. Dubbing may be used for pullets kept in cages to prevent injury to the combs from the wire. Detoeing is used on market turkeys to prevent injury from the birds walking on

FIG. 5.7.   Chicks hatching in an incubator. Source: USDA.

each other when startled and scratching each other. Each of these services increases the price of a chick. The final hatchery service is the safe and sanitary delivery of the newly hatched, vigorous, healthy chicks (Fig. 5.7).

REFERENCES

Anonymous. 1976. From Egg to Chick, a 4-H Manual of Embryology and Incubation. NE Coop. Pub./NE, University of Maryland, College Park, MD.

Romanoff, A. L., and Romanoff, A. J. 1949. The Avian Egg. John Wiley, New York.

Romanoff, A. L., and Romanoff, A. J. 1967. Biochemistry of the Avian Embryo. Interscience Publishers, New York.

6

Social Behavior and
Animal Welfare

Although man has studied animal behavior and utilized this knowledge to exploit animals since he first started hunting them for food in prehistoric times, the study of animal behavior is one of the newer scientific disciplines. Poultry producers and housing and equipment designers need to be aware of the *behavioral* characteristics of birds so that meat and eggs can be produced with optimum efficiency and with concern for the welfare and well-being of the birds.

Because poultry are social animals, they live in flocks or groups. To live with one another and offer mutual protection, they communicate by using voice, actions, and displays of plumage or adornments. Considerable research has been conducted to interpret avian behavior and the factors that motivate birds to react to the voices, actions, and displays of other members of the species.

SOCIAL ORGANIZATION

All groups of birds tend to develop a social order. Even people with little knowledge of chickens are aware of the order of rank and dominance in chickens commonly referred to as the "peck order." Thorleif Schjelderup-Ebbe, a Norwegian scientist, was one of the pioneers in studying the social organizations of animals that live in groups. He observed that peck orders in chickens tend to become increasingly complex as group size increases. When both sexes are present, there are actually two peck orders, one for males and one for females. Peck orders for the two sexes are separate and unequal because all hens give way to all cocks (Fig. 6.1).

In the complex peck order (Fig. 6.1), the dominant hen (A) can peck all others but none dares peck her. The number two hen (B) was pecked by (A) and yielded to her, but (B) could peck all other hens and dominate them. The third hen (C) was pecked by (A) and (B) but could peck (D), (E), (F), etc. The hen at the bottom right was pecked by all other hens but could not fight back.

Peck orders are useful in reducing conflicts in a group. After rank has been established, lower ranking members yield to threats and true fighting is avoided. Peck order appears to be related to an individual's need for some space of his or her own. This space for chickens is primarily in front of the face. For example, cocks are prepared for fighting by being held face to face with other cocks to elicit antagonistic behavior.

Scientists have classified bird behavior into actions created by eight general stimuli:

1. Limited resources (aggression, competition, searching)
2. Food
3. Physiological needs (food, rest, warmth)
4. Parent–offspring relationships
5. Sexual relationships
6. Self-care needs (preening, comfort movements)
7. Novel objects and environments (investigation, avoidance, immobility)
8. Strange birds (investigation, aggressive or submissive acts).

BEHAVIOR OF FERAL CHICKENS

The behavior, communications, and reactions of chickens to these eight stimuli can best be illustrated by McBride's study of feral chickens (McBride et al., 1969).

Humans abandoned an island off the coast of Australia in 1928 but left the chickens on the island to forage for themselves. The estimated poultry population increased to as high as 1500 birds in late autumn

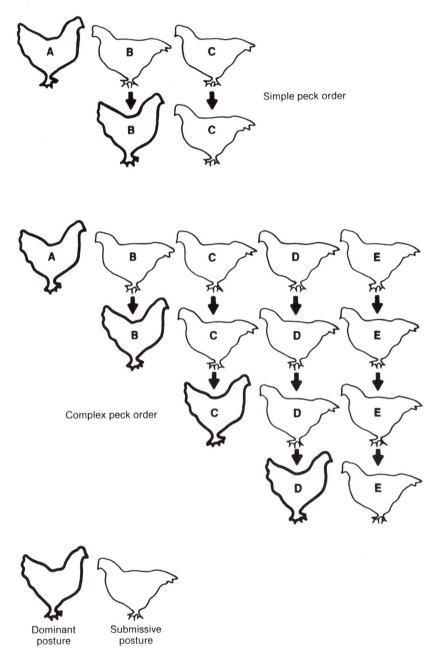

FIG. 6.1. Peck orders.

and declined to as few as 600 birds in the spring. Predators, primarily cats and hunters, were the main causes of mortality.

The chickens became seasonal breeders again with each hen hatching one or two broods per year. During the breeding season each cock established a territory with one or usually several females. Each group roosted in trees near the center of a home range or territory and each bird tended to roost in the same place each night. The distance between roosting trees for dominant males and groups of females varied from 200 ft (60 m) in dense tree areas to 500 ft (150 m) in sparse tree areas. Each dominant male patrolled the boundaries of his area daily. Two neighboring males would often approach the boundary and walk along the territorial line 50 ft (15 m) apart and crow. Males with smaller numbers of females in their harem were most active in boundary patrol. There were many other ramifications of territory control in these feral birds and overlapping of territories also occurred.

COMMUNICATION

A number of methods of communication occurred, both visual and auditory. Visually posture was one method observed: tail carriage was important, both spreading, raising, and lowering, as well as hackles and wing carriage. Sounds were another method of communication. These were subdivided into (1) cackles, (2) clucks, (3) chirps, (4) crows, and (5) squawks.

Cackles consisted of a series of clucklike sounds with occasional louder squawks. When a predator was on the ground, cackles consisted of a rapid series of clucks, by either males or females, that served as an alert. An alarm call, which was a type of squawk, signaled the flock to hide or remain motionless. A disturbance cackle was also used after the hen left the nest following egg laying, possibly as a decoy to aid in hiding the nest. Chickens could distinguish between the alarm cackle and the cackle of a laying hen even though they sounded similar.

Clucks were used mainly by males and broody hens. When clucks were emitted in rapid succession they were an alert or attention call used by the male when tidbitting, during courtship, followed by waltzing. Broody hens used clucks as a location call for their chicks.

Chirps or twitters were used for contact, for location, and for distress calls by the chicks.

The crow was used by males in a variety of situations, with each male's crow being quite distinct from that of all other males.

The crow was used by each male to identify himself to other males. Crowing would start in bursts as early as 2 hr before sunrise and become almost continuous just before dawn.

Territorial crows were heard before dawn at the roosting site, and at more or less regular positions while males were patrolling their territories during the daytime. Feral males did much of this patrolling of territory on foot.

Status crows were used to maintain and assert social rank and usually followed a wing flap or occurred after a dominance interaction.

Location crows were used by males as they moved about their territory to keep each male and his hens separate from their neighboring group. Also, these location crows had threat functions, whereas the location squawk involved no threat.

"All clear" crows were sounded by males after danger had passed.

Mating calls were given in early morning and mating with each hen occurred as she descended from the roost.

A group assembly call was used to gather the group together while a roosting call was given by the dominant male after he had taken his roosting location for the night.

Squawks had several meanings. Three of these were general to adult males and females, as well as to juveniles.

An aerial predator call was a warning consisting of a squawk that rose and fell slowly and was taken up by others in the group who immediately stood or squatted motionless. When an attack was made by an aerial predator a fighting posture was displayed combined with a loud squawk.

A distress squawk was given by a subordinate bird when avoiding a sudden attack by another bird.

Location squawks were gentler sounds consisting of a long drawn squawk followed by a series of shorter ones.

The egg call consisted of squawking more often, prior to laying, when the hen left the group and approached her nest. It was usually rather loud with the beak opened. Domestic hens often give this call while watching the attendant approach the pen.

In addition to these calls and squawks, the broody hen rearing chicks has a very elaborate system of communicating with her chicks and vice versa. These change as the chicks age until they are finally weaned from the mother's care and disowned insofar as family relations are concerned.

BEHAVIOR OF DOMESTIC POULTRY

Changes in domestic behavior when territory is limited to a small yard, as compared with a large territory, indicate modifications that become extreme in large mated flocks. For this reason, they are most difficult to study. In large flocks of domestic hens, each hen moves in a limited area within the house. The further she moves from the center of her area, the more pecks she receives and the fewer she gives. Thus, the domestic hen seems to be more dominant at the center of her home range and the pecking she receives appears to provide a mechanism for keeping her within this restricted area. The dominance system is based on peck right rather than peck dominance. The essence of social organization appears to be communication and behavior which affect spacing between individuals. The spacing that is most important is that in front of the face. Tidbitting of males may be a spacing factor if the tail is raised and a feeding attractor if the tail is lowered. When the tail is raised, it is a sign for the intruder to move out. Although tidbitting is used by the male in courtship, it is never followed by copulation. It has been difficult to separate those aspects of social behavior that are learned from those that are innate. Imprinting, for example, involves innate following of a conspicuous object, which typically leads to social bonding of the baby chick to its mother, when hens are allowed to hatch their own chicks. Baby chicks also bond to other chicks, particularly when no mother hen is present, as in incubator hatched chicks. Learning is involved in determining what the chick forms a bond with. Peck orders and the other patterns develop later during the growing period.

SOCIAL BEHAVIOR IN GROWING CHICKS

During the first month, chicks may wander as far as 60 ft (18 m) from other chicks or their brooding mother, but they regularly return to close visual contact. This varies from 10 to 15 ft (3–5 m) while the chicks are quite young, up to about 25 ft (8 m) for month-old chicks. The chicks return to closer distances more frequently when very young. One chick running toward another chick or its dam is a stimulus for other chicks to join the group.

Up to 6 weeks of age social behavior consists of companionship. At about 6 weeks of age agonistic behavior gradually begins. Prior to this time there is a pattern of play behavior that mimics agonistic behavior. Play is initiated by a chick running an irregular course, usually circular, with wings flapping, then running up to another chick and adopting the male fighting posture, neck outstretched with hackle feathers raised. At about 6 weeks of age this play develops into a mild dominance fight. Each round of a dominance fight ends with avoidance and motionless postures. The bird that loses starts to avoid and then stands motionless. The dominating bird also stands motionless but is always the first to move away, even if only a few inches, to return and stand motionless again. If the submitting bird moves before the dominant one, the fight starts again and may comprise several rounds before ending with the motionless submission.

The first attempts at mating are observed at about 6–8 weeks of age. When hens are brooding chicks, this first mating reaction by male chicks is directed curiously toward their mother. In artificially brooded chicks, the mating behavior of males is directed toward the female chicks. Chicks in the feral flocks were normally "weaned" from their mothers care at about 10–12 weeks of age.

The beginning of crowing in male chicks varies with the degree of sexual development of the strain. In feral birds it may be as late as 17 weeks. In broiler chicks it may occur as early as 8 or 9 weeks after hatching.

Streaming is a little understood behavioral phenomenon in young chicks. It consists of one bird running across the pen with most of the rest joining in a group run toward the same point. In feral groups brooded by different hens, this appears to flock the groups into one. Chicks are always attracted to running birds of their own species. It seems likely that the "streaming" behavior pattern may be related to "fowl hysteria" or "piling up" of younger birds when frightened. Streaming in feral settings may group birds so that the incidence of inbreeding is minimized.

OTHER ADULT SOCIAL BEHAVIOR CHARACTERISTICS

Peck order among sexually mature males is expressed as a spatial arrangement. The dominant male controls the behavior of the next subordinate male which tends to remain nearest to him. These second order males in turn control the behavior of their subordinates. All distances are reduced as dusk approaches. This is why in domestic poultry lowered light levels tend to have a calming effect on birds, limiting activity by mimicking approaching darkness.

Chickens never move randomly with regard to their neighbors. Dominant individuals face their subordinates, whereas the subordinates avoid facing a dominant individual. Social rank is constantly being recognized and regulates behavior without leading to any overt aggressiveness. Males express status while standing or walking. All chickens must constantly maintain their rank in the peck order by appropriate behavior. Loss of status is best exemplified by running, which is a submissive behavior. After a male has run, it will stop, stand high on its toes, ruffle its feathers, flap its wings, and usually crow. This will occur in both birds when a dominant male has chased a subordinate one. Sudden changes in posture associated with mating behavior of the cock may lead to temporary loss of recognition by subordinates, which then results in interference. In any case the result is of very brief duration, decreases with experience, and tends to be without significance in terms of mating effectiveness by dominant males. If a male submits in the presence of a dominant male, its feathers are compressed and its tail is lowered and closed. Walking the appropriate distance from the dominant male expresses status. Males always stand facing each other, always sideways. The normal distance between a dominant male and his subordinate in which status and thrust behavior are avoided is between 15 and 20 ft (5–6 m), although it is reduced at night in roosting areas.

All types of social behavior are subject to communication, spacing, and dominance. This is complicated by the interaction of subgroups within a large pen. Each dominant male carries with him a spatial control, 8–10 ft (3 m) in radius for females and about 15 ft (5 m) in radius with subordinate males. In this area, the dominant male exhibits all dominance and status behavior.

McBride thinks that the key criterion in animal organization is spacing and its regulation in regard to the following:

1. Fixed territories
2. Overlapping territories with fixed dominance
3. Overlapping territories with dominance a function of distance from territory centers
4. Home ranges with personal spheres in which the animal defends a two- or three-dimensional area about it
5. A social dominance hierarchy in which the personal sphere is reduced to an area in front of the face

ANIMAL RIGHTS

Concern for animal well-being has always existed, but during the past several decades there has been a growing movement for animal rights. In the last 20–30 years an incredibly small portion of the population of developed countries has been able to supply food to the rest of the population. Modern agriculture became efficient, producing food at low cost desirable to the nonfarming public; this was accomplished by intensive farming practices including confinement and high-density livestock production. Production practices also changed rapidly by substituting capital for labor so that modern farming now involves mass production of commodities such as eggs and broilers. With technology allowing fewer farmers to produce the nation's food items, the general public has become further removed from agriculture. There are now whole generations of people who have not harvested food items, particularly of an animal source.

Some groups of consumers have become disturbed about the animal production methods used to produce food economically and have chosen to voice their views politically, philosophically, and, in several instances, violently. Farmers are just beginning to become aware of the concerns of the humane movement and its implications. Efforts by farmers to react to the problem have ranged from an early defensive posture to private soul searching and the initiation of guidelines for "humane" production standards by some segments of agriculture.

THE HUMANE MOVEMENT

The animal humane movement is fostered by a variety of organizations, none of whom speaks for the movement as a whole. Lack of a central agency speaking for the movement has made it difficult for agricultural interests to communicate effectively with these groups because the movement is made up of individuals with widely different views, beliefs, and motivations. Overall, it appears that there are three philosophies involved. They are animal welfarists, animal rightists, and animal liberationists.

ANIMAL WELFARISTS

Animal welfarists believe that animals should be treated humanely. The problem lies in the perception of what is humane. Humane treatment varies from concern about absence of gross cruelty (beating, starvation, mutilation) to concern about psychological deprivation of individual animals (lack of social contact, play activity, exposure to a natural environment). This group, in most cases, accepts the use of animals in agriculture, but may decry particular production practices, for example, caging layers. Communication with animal welfarists by agricultural groups should be encouraged because it is an educational process for all people involved. On the one hand, members of the agricultural community must realize that some practices are objectionable and welfarists on the other hand need to increase their knowledge of modern agriculture, its benefits, and the price to be paid for change.

ANIMAL RIGHTISTS

Animal rightists believe that animals have inalienable rights, similar to humans, and that mankind does not have exclusive dominion over animals to exploit them. Modern agricultural practices are viewed as slavery at best by this group. Abolition of current agricultural practices is viewed as a long-term goal. Rightists views are generally unalterable, supported by moral, philosophical, or religious tenets, not susceptible to direct argument. The rightist group often advances technically impossible solutions and appears to be prone to sensationalism.

ANIMAL LIBERATIONISTS

Animal liberationists desire the overnight abolition of any exploitation of animals in any form, including zoos, confinement agriculture, and research laboratories. A few members of these groups are the potentially violent members of the humane movement, unwilling to wait or work for peaceful change. Propaganda is a standard technique utilized by this small and restless group.

AGRICULTURAL INVOLVEMENT WITH HUMANE GROUPS

Commodity groups, such as the various segments of the poultry industry, are just beginning to develop and enforce codes of ethical practice for animal production and to notify the public of these activities. It is essential that the humane movement and the general public realize that agriculture "cares" about the welfare of its animals and tries to prevent abuse before it occurs. Most cruelty cases involving farm animals involve negligence or sadistic behavior by a few individuals. The main body of agriculture should divorce itself from these individuals.

PUBLIC RELATIONS

It is important for farmers to maintain a free exchange of ideas and information and help educate the general public. Tours of facilities and exchanges of ideas between farmers and nonfarmers should be encouraged and industry economics and production problems should be discussed. Agricultural groups and individuals need to explain to the public what is happening in agriculture, since the general public has little perception of the complexity of modern food animal production and very little understanding of the consequences of overregulation.

Farmers have a mission to feed people by producing feed efficiently, but if the general public truly desires less chicken or "humanely produced" chicken, or thinks producing no chicken will solve starvation in the world, agriculture has no choice but to comply. On the other hand, farmers have to help ensure that the public makes an informed decision that will not destabilize our society.

NEED FOR A CODE OF ETHICS

To establish codes of ethics, one must first accept certain basic facts. Biologically, man is omnivorous and at the top of the food chain. He has always been a predator who gathered and ate plants and harvested animals and animal products for his own survival and well-being. This practice of systematically exploiting plants and animals for his own benefit is now called agriculture.

In the area of animal welfare, there is some general agreement on the extreme limits of acceptable and unacceptable humane practices; for example, unnecessary suffering and cruelty are not condoned. Unfortunately, some current animal husbandry practices are considered to be in a gray area.

A few examples of poultry husbandry practices that have been questioned include the following:

1. Disposal of hatchery wastes including live and still hatching baby chicks
2. Caging growing pullets or laying hens
3. Some practices used in catching and transporting broilers
4. Slaughter procedures
5. Depriving birds of food and water to bring about forced molting
6. Lack of sufficient space for birds grown on the floor
7. Beak trimming, dubbing, dewinging, detoeing, and other hatchery-initiated management practices

As a result of the many questions raised by debate on this subject, considerable research is being initiated to measure animals' physical and mental discomfort under varied conditions. Several indicators of poultry welfare have been investigated and the field of investigation continues to be an active one. Eventually the issue must be resolved by all segments of society by developing codes of ethics. Hopefully, such decisions will be based on facts and not on emotions.

The humane animal movement is growing and more legislation involving animal production practices will result in the near future. For this reason, the humane animal movement cannot be ignored. Poultrymen will have to take animal welfare into consideration in planning future operations much more than they have in the past.

REFERENCES

Craig, J. V. 1981. Domestic Animal Behavior: Causes and Implications for Animal Care and Management. Prentice-Hall, Englewood Cliffs, NJ.

Craig, J. V., and Adams, A. W. 1984. Behavior and well-being of hens *(Gallus domesticus)* in alternative housing environments. World's Poultry Sci. J. *40*, 221–240.

Guhl, A. M. 1962. The behavior of chickens. *In* The Behavior of Domestic Animals. E. S. E. Hafez (Editor). Bailliere, Tindall and Cox, London.

Limburg, P. R. 1975. Chickens, Chickens, Chickens. Thomas Nelson, Nashville, TN.

McBride, G., and Foenander, F. 1961. Territorial behavior in the domestic fowl. Nature (London) *194*, 192.

McBride, G., Parer, I. P., and Foenander, F. 1969. The social organization and behavior of the domestic fowl. Animal Behavior Monographs, Part 3, pp. 126–181. Baillere, Tindall and Cassell, London.

Regan, T., and Singer, P. 1976. Animal Rights and Human Obligations. Prentice-Hall, Englewood Cliffs, NJ.

Wood-Gush, D. G. M. 1955. The behavior of the domestic chicken: A review of the literature. Br. J. Anim. Behav. *3*, 81–110.

7

Environment and Housing

Housing performs a number of functions. Among the more obvious are protection from heat, cold, rain, snow, sleet, wind, and extremes of noise, light, and darkness. Houses also protect birds from predators and theft and make it easier to control diseases and parasites. From a management point of view houses keep birds assembled together in one unit in which the manager handles the birds as a single group. All birds are subjected to the same environment, feed, water, medication, and management.

House design and construction vary considerably depending on the climate and area of the country (Fig. 7.1). For example, in northern climates the foundations must be heavier and deeper in the ground and the roof trusses heavier to withstand snowloads. Heavier insulation is also required. Despite these different requirements, a typical house is generally 36–40 ft wide with the length varying from 360 to 500 ft long. Older houses often were built with a headhouse or service area in the middle with wings off each end. The need for service areas has been greatly reduced by bulk feed handling and automated feeding systems, egg gathering belts, and other labor-saving devices. Some new environmental houses have no windows, which makes them completely dependent on electricity to run fans and lights.

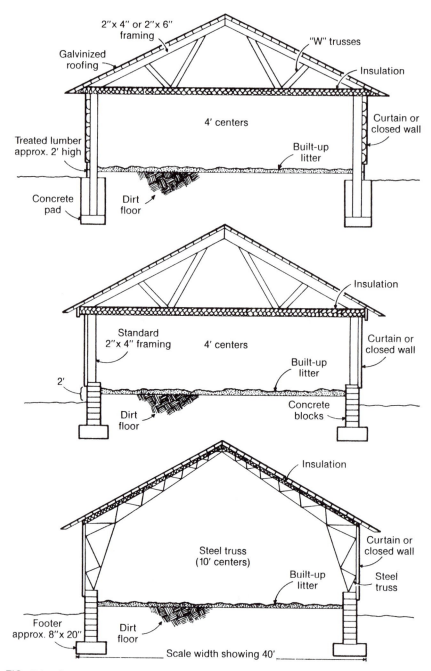

FIG. 7.1. Cross sections of the structures of typical modern poultry houses. Generally houses are 40 ft long and whatever length is desired. Depending on the interior fixtures, the houses can be used for brooding, broiler, or turkey production, housing breeders, or pullet and laying cages. Top, pole construction; center, conventional foundation, 2 × 4 ft frame, and truss construction with foundation often used in colder climates; bottom, a prefabricated steel truss design.

FARMSTEAD PLANNING

Several criteria are important in planning a poultry enterprise. Poultry farms should be located far from housing developments or any land that is planned for a subdivision. Unfortunately, although a poultry farm may have been in operation for many years, if a housing development starts near a poultry farm the human occupants of the houses are usually favored. Most counties now have zoning and building codes. For this reason, it is well to check and comply with all requirements before starting any construction. Checking prevents costly alterations and delays during later construction.

Poultry farms should also be planned to allow easy movement of vehicles such as trucks used for bulk feed deliveries and trailer trucks used for loading poultry. Because the poultry farm is home for most families, it should be as pleasant and attractive as possible. This does not necessarily mean elaborate or expensive construction but surroundings that are well planned and usable. Because of economic constraints, the poultry operation is often given consideration over family needs. Most poultrymen need to be more aware of aesthetics, labor efficiency, and environmental conditions, particularly those related to pollution and nuisance complaints. To do good work and remain productive, poultry farmers and their families must be happy and content. The farmstead plan should also take into account safety precautions for the family. This includes access for vehicles as they reach main roads, fire protection for all buildings, and the safety of children, who in playing may run in and out of farm drives or play near equipment such as farm ponds, lagoons, and other potentially dangerous areas.

EXTERNAL SERVICES AND UTILITIES

Close proximity to hard surface roads is essential because poultry farming operations depend on motor transportation. Service roads into farms from hard surface roads should be well planned, with at least 16 ft of clearance and 6 ft of graveled or paved driving surface. The initial cost of a makeshift service road may seem reasonable but the maintenance over the life of the buildings may be considerable and should be taken into account in the initial construction plans. Because poultry operations use large amounts of electricity, plans should include provisions for an ample power supply. As examples, distribution should be from one central pole so that the source is adequate, safe, and economical and three-phase power should be used when possible to reduce the current requirements of electric motors. Since poultrymen are dependent on electricity to ventilate, feed, and water their livestock, those areas of the country suffering problems with high winds, heavy snow, or ice storms should bury their supply lines in a trench. A self-contained or tractor-operated auxiliary generator should also be available. A portable unit is a safety precaution so that, if the need arises, water pumps, automatic feeders, and ventilation fans can be operated for short periods to keep the birds comfortable. Modern farm planners also take advantage of the peak load demand rate charges made by utility companies by using as much electricity as possible during off peak periods.

WATER SUPPLY

The presence of a sufficient water supply of adequate quality should be confirmed before the poultry house site is prepared for construction. For example, regulations from local health departments should be checked so that the wells and the distribution system meet required specifications. One central well system usually is best because such a system reduces friction loss in pipes and leaves open the opportunity for expansion. A 1-in. pipe is a minimum for distribution lines that should be buried deep enough to prevent freezing.

On large poultry farms drilling two or more wells and cross-connecting them to provide an uninterrupted supply of water when problems occur with pumps and motors is also a common practice. Water systems should include a method for easily switching to a standby electrical source in case of power failure. With the water lines having adequate and easily accessible cutoffs,' water meters should also be used to measure water consumption, which is an important indicator of flock health. Water lines entering houses should also be plumbed so medicators can be installed.

POULTRY BUILDING ORIENTATION

In general, houses should be located away from nuisances such as highways, railroad tracks, and airports. They should also be located to catch prevailing breezes and downwind from any dwellings. Most of the conventional curtain-sided houses built for broilers are built running east and west to help control the heat in summer. Many houses have been built to follow the way of the land with little thought given to natural ventilation. Another consideration in site selection is the possibility of the future use of solar collectors. To make optimum use of solar collectors buildings need a southern exposure. Buildings should also be spaced at least 50 ft or more apart to obtain low fire insurance rates and to reduce the risk of multiple building fires but close enough to the house so the birds can be easily checked at night in bad weather.

Poultry house sites should be properly graded to elevate the site and direct surface water away from the buildings. Many sites are inadequately prepared before the houses are built. Unfortunately, some of these problems can never be adequately remedied after the house is constructed. A modern poultry house can easily have over a half-acre of space under the roof and during heavy rain the run-off between houses is tremendous. Grass waterways can be utilized to prevent erosion and are attractive yet easy to maintain.

AESTHETIC VALUE

Poultry farms should be landscaped the same as other buildings. Landscaping does not have to be elaborate or expensive. All areas should be seeded with a low maintenance grass or other suitable ground cover. Plantings can be used to shield undesirable features such as incinerators, trash cans, and other less desirable areas and for wind breaks (Fig. 7.2).

FIG. 7.2. A well-designed broiler farm layout.

HOUSE CONSTRUCTION

Poultry houses are built to provide as much comfort as possible for the birds while keeping the costs and labor requirements as low as possible. Houses must pay for themselves and provide a profit. Although construction costs must be the main consideration, material quality and craftsmanship should be adequate to build a satisfactory house that does not undergo undue depreciation and have high maintenance costs.

TURNKEY HOUSES

Most poultry houses built today are factory-built or turnkey houses that come complete with insulation, a ventilating system, all equipment, and the house ready for birds when the builder leaves the site. One important advantage of this arrangement is that the price is agreed upon in advance and the house is built to certain specifications. Financing the house is often easier with this type of construction because similar houses have usually been financed in the area. Since most poultry today is produced under contract, with the housing requirements specified by the contractor and the design, and the quality of materials and the labor crew are often under contract to the firm contracting the individual's production, both the producer and the financier know the type of house to expect. Although some independent poultry producers still build their own houses, the size of modern day houses makes building one a major construction project.

CRITERIA FOR SELECTING A MANUFACTURED HOUSE

Some considerations that should be given to poultry house construction when building or inspecting factory-built houses are as follows.

Foundations

Poultry houses generally are constructed using a concrete footer with at least three courses of concrete blocks laid on the footer. Foundations must be adequate to support the building, yet deep enough to prevent heaving by freezing temperatures and high enough to keep out surface water.

Floors

Floors in modern poultry houses are of three types. Broilers and turkeys reared for meat purposes are usually reared on dirt floors covered with litter and cleaned each year. Dirt floors can be sprayed to reduce the number of disease organisms present. At this time concrete floors in growout operations are not feasible because of the cost.

Slat floors are popular for some phases of poultry production such as broiler breeders. The slats are generally placed over two-thirds of the floor area with the feeders and waterers located on the slats. The other third is a dirt floor on which the birds can more easily mate. Such a system provides good fertility and keeps hatching eggs cleaner than with other systems.

Concrete floors are widely used in caged layer houses for commercial egg production. Such floors are moisture proof, rat proof, durable, and sanitary and in areas with cold temperatures, concrete floors help prevent wide fluctuations in temperatures.

Walls and Ends

The walls and ends of houses should be of solid enough construction to support the roof and withstand heavy wind and a possible snow load. Considerable variation is possible in the use of building materials, depending on cost, availability, and the degree of insulation available. Modern day poultry houses do not have partitions and if dividers are placed in houses they are constructed of a wire frame and are not load bearing. However, the use of masonry construction for the walls of a central headhouse to serve as a fire stop is a good practice.

Trusses

Modern poultry houses are constructed with gable roofs. Ideally, these houses are built to provide a clear span with up to 40 ft (13 m). In widths in excess of 40 ft, support posts are utilized because of the high costs for this span of trusses. Trusses can be constructed of wood or metal with the truss spacing dependent on design and load requirements (Fig. 7.3). Trusses in cage laying houses need reinforcement to support the weight of the birds and cages.

Roofs

Almost all poultry houses constructed today use metal roofs of either aluminum or galvanized steel. These materials are popular because of

FIG. 7.3. Interior of a steel truss, clear span broiler house.

their low maintenance requirements and durability. In addition, the widths and lengths of the roofing materials make them quick and easy to install, which results in saving a considerable amount of labor. Several new roofing materials have recently become available that should work well for poultry house roofs and even ends and sidewalls. These materials are rigid formed corrugated asphalt sheets which are expected to last 25 years. Poultry house roofs should also be insulated during construction.

Insulation

The term heat loss refers to the movement of heat from warmer to colder areas or surfaces. Heat can be lost from a poultry building through the building's materials, air spaces, cracks, doors, or curtain openings. Insulation does not stop the passage of heat but merely slows the rate of heat movement. All poultry houses constructed today are insulated, at least in the roof, and many are now being insulated in the ends and sidewalls as well. Roof insulation aids bird comfort both in the summer and winter by helping to reduce temperature extremes. Insulation values are shown for various materials in Table 7.1. Insulation materials have a high resistance to the flow of heat. Most are fibrous or granular materials that contain many air pockets or cells. The more trapped air pockets in a material, the better its value as insulation. Some materials, such as wood, have good insulating values

TABLE 7.1. *R* Values of Various Building Materials

Item	Thickness		Resistance
	In.	Cm	Rating
Insulation per 1 in. (2.5 cm) of thickness			
Blanket bat	1	2.5	3.70
Balsam wool (wood fiber blanket)	1	2.5	4.00
Cellulose fiber	1	2.5	4.16
Expanded polystyrene, molded (bead board)	1	2.5	3.50
Expanded polystyrene, extruded (Styrofoam)	1	2.5	5.00
Urethane foam	1	2.5	6.60
Fiberglass (glass wool)	1	2.5	3.70
Palco wool (redwood fiber)	1	2.5	3.84
Rock wool (machine blown)	1	2.5	3.33
Rock wool (blanket)	1	2.5	3.33
Foam glass	1	2.5	2.50
Glass fiber blanket	1	2.5	3.33
Mineral wool	1	2.5	3.33
Insulation board	1	2.5	0.00
Vermiculite (expanded)	1	2.5	2.05
Wood fiber	1	2.5	3.33
Sawdust or shavings (dry)	1	2.5	2.22
Straw	1	2.5	1.75
Materials (thickness as indicated)			
Air space, horizontal	0.75+	1.8+	2.33
Air space, vertical	0.75+	1.8+	0.91
Asbestos cement	0.12	0.3	0.03
Building paper			0.15
Concrete	8.00	20.3	0.61
Concrete block	8.00	20.3	1.11
Hardboard	0.25	0.6	0.18
Plywood	0.25	0.6	0.32
Plywood	0.50	1.2	0.63
Surface, inside			0.61
Surface, outside			0.17
Siding, drop	0.75	1.9	0.94
Sheathing	0.75	1.9	0.92
Metal siding			0.09
Glass, single			0.61
Shingles, asbestos			0.18
Shingles, wood			0.78
Roofing (roll, 55-lb)			0.15
Vapor barrier			0.15

Source: North (1978).

while others such as concrete are very poor insulators. Table 7.1 lists the *R* values for selected building materials. For example, 1 in. of wood has approximately 10 times the insulating value of concrete. Popular insulation types used in poultry houses are blanket insulation, batt insulation, and various formed sheets of the foam insulation type. Spray-on foam insulation works well when remodeling older uninsulated houses. Rising brooding fuel costs and electrical costs associated with fan ventilation have forced poultrymen to consider the economic feasibility of better insulated houses. Insulation in poultry houses should provide for an R value of 12–15 in ceilings and 8–10 in the ends and sidewalls. Insulated curtains or narrow curtains with insulated inserts are being used in some new houses.

Loss of heat in poultry buildings can occur in one of the following ways or combinations of ways. Heat movement by conduction occurs whenever there is direct contact between hot and cold areas. If one side of a material is heated, the heat will be conducted through the material to the cooler side. Heat movement by convection usually depends on air or water to convey heat from the warm to the colder areas.

Heat movement by radiation occurs when two surfaces are at different temperatures. The warmer body or surface will radiate heat to the cooler surface without heating the air between. The sun's heat is an example of radiated heat. Poultry can lose heat by radiation to cold surfaces such as walls or ceilings. Insulation reduces the heat loss from a building by any one of these methods.

VAPOR BARRIERS

Vapor barriers are used to prevent condensation of moisture in the walls, ceiling, and insulation. At some theoretical point in the insulation the outside cold air contacts and cools the warmer inside air which causes condensation and wet insulation. To prevent this from occurring, a waterproof barrier is installed on the inside walls of the house. Vapor barriers can be Kraft paper attached directly to the insulation or a liner of plastic insulation. The vapor barrier is always placed on the warm side of the wall which contains the higher moisture content.

VENTILATION

The physiological structure of poultry differs from mammals in many ways. Of particular importance is the fact that urine is excreted in solid form, which means moisture must be discarded from the body through the respiratory process. This means that poultry houses must be designed differently from other farm animal shelters. If the warm moisture-laden air given off by a poultry flock is not removed from the house before cooling, the moisture condenses to cause moisture, ammonia, and litter problems in the house. Litter cannot absorb all the moisture.

Because of the development of high-density poultry housing systems, it is now economically feasible to use high-velocity electrically powered fans for ventilation. The ultimate is the environmentally controlled house. However, most poultry houses are still constructed with open sides and curtains that can be adjusted according to climatic conditions. Some housing types also contain some sort of roof ventilation. Other houses use a combination of fans and natural ventilation.

As mentioned earlier, the high body temperature of poultry coupled with the inability of poultry to sweat means that the respiratory system must be used to dissipate heat in the form of water vapor.

In modern poultry operations today, there are basically two types of ventilation systems: forced air or a type of gravity convection.

Forced-Air Systems

A forced-air ventilation system utilizes fans to provide conditions for positive- or negative-pressure ventilation (Fig. 7.4). These systems allow a mixture of air and provide for a better air exchange compared to when air is simply exhausted without pressure buildup.

Many older poultry houses used a gravity convection system in which the warm moisture-laden air rises through an opening or ridge ventilator at the top of the house causing a pressure differential at bird level that allows fresh air to come in along the sides. Adjustments are made by raising or lowering the curtains covering the openings. In this system fans could be used to stir the air inside the house (Fig. 7.5) and this system has been used in broiler houses in the south.

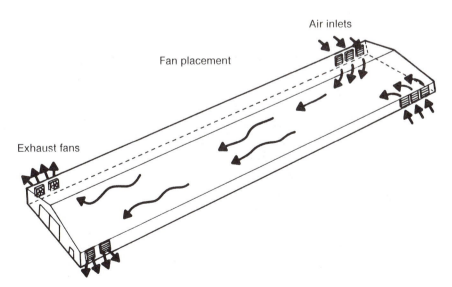

FIG. 7.4. Example of a negative-pressure forced-air ventilating system.

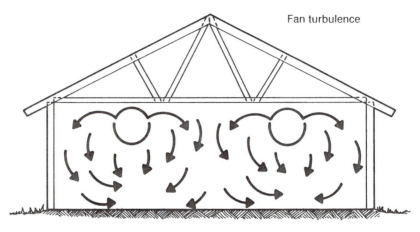

FIG. 7.5.   Fan turbulence to ensure uniform air circulation.

Negative-Pressure Systems

Negative-pressure ventilation systems utilize banks of fans along one side of the building that exhaust air causing a negative pressure in the house. A baffle system located under the eaves with an adjustable slat opening is used to control the flow of fresh air into the house. This system can cause some draft problems if not properly operated. The rate of air movement in this type of house varies with the climate, housing type, bird size, and density. In broiler houses (Fig. 7.6) the

FIG. 7.6.   A broiler house with a negative-pressure fan system.

capacity for air movement should be at least 4 ft^3 of air per bird a minute. Preferably 2–8 ft^3 (0.06–0.23 m^3) of air per minute per bird fan capacity would be realistic. Less capacity could be used in cool climates. Cool weather ventilation plans should utilize variable-speed fans so that cold air is not drawn in on the birds.

Positive-Pressure Systems

Positive-pressure systems of several types have been developed. In layer houses, fans located on the end of the house push air into suspended duct tubes running the length of the house. Small holes in the tube allow the air to escape and mix with the house air. When the air pressure in the house builds up the warm air can be forced out an opening near floor level through an exhaust stack located at the end of the house. Static pressure controllers are used to regulate air inlets and maintain a positive pressure in growout houses for broilers and turkeys.

REFERENCES

Anonymous. 1982. Farmstead Planning Handbook. Midwest Plan Service. Iowa State University, Ames, IA.

Anonymous. 1984. Designs for Glued Trusses. Midwest Plan Service. Iowa State University, Ames, IA.

Barre, H. J., Sammett, L. L., and Nelson, G. L. 1987. Environmental and Functional Engineering for Agricultural Buildings. AVI Publishing Company, Westport, CT, in press.

Esmay, M. L., and Dixon, J. E. 1986. Environmental Control for Agricultural Buildings. AVI Publishing Company, Westport, CT.

Esmay, M. L. 1960. Design analysis for poultry-house ventilation. J. Agr. Eng. *41*, 9.

North, M. O. 1978. Commercial Chicken Production Manual, 2nd Edition. AVI Publishing Company, Westport, CT.

8

Poultry Nutrition

In the United States the economic value of poultry feed is important. From 35 to 45% of all feed produced by the commercial feed industry is for poultry and from 50 to 75% of the cost of poultry production is for feed.

Poultry in their natural habitat are omnivores. They eat both plants and animals. Since the domestication of poultry, extensive research has been conducted on their nutritional needs and food requirements (proteins, carbohydrates, fats, vitamins, and minerals). Commercially prepared rations are calculated and mixed with the aid of a computer to obtain the lowest cost ingredients for an optimum balanced ration.

Although the nutritional needs of poultry are better known than for any other animal including humans, there is still considerable variability in their requirements. This is because needs vary for different species, ages, sexes, productive status, genetic background, environmental temperatures, housing conditions, stresses, and management systems, and because of variations in quality in the same lots of feed ingredients, nutrient interrelationships, improper mixing, and a number of other variables.

Despite these limitations most commercial poultry are raised to grow and reproduce throughout their entire lifetime in total confinement. The only feed they receive is provided by the poultryman. (Feed storage bins in a modern poultry operation are shown in Fig. 8.1.)

The main objective of the poultry producer is to achieve efficient economic conversion of feedstuffs into human food. To achieve this it is necessary to recognize that feed must be used for needs other than the production of muscle, organ tissue, and eggs. Any food consumed is used first to provide the energy necessary to maintain a normal body temperature, to provide the body movements necessary to survive, and to provide energy for the production of body secretions and the repair of body tissues, and second for growth and finally for reproduction. Any additional nutrients are stored in the bird as fat. Growth involves the development of new bone, muscle (meat), skin, nerves, organs, and feathers.

Poultry are more sensitive to nutrient needs and balance than mammals since their body and digestive processes are more rapid, their respiration and circulation are faster, and their body temperature is 8–10° higher (42°C, 107°F). For these reasons growth takes place at a rapid rate and poultry mature at an early age. Because their rapid growth makes them sensitive to nutritional deficiencies, poultry are frequently used as laboratory animals in nutrition experiments; as a result our knowledge of poultry nutrition is probably more advanced than for any other animal.

FIG. 8.1.   Feed storage bins in a modern poultry operation. Feed is conveyed directly from the trucks to the bins and then by automatic feeders to the chickens.

NUTRIENTS

Poultry require almost the same nutrients—water, carbohydrates, fats, proteins, minerals, vitamins, and certain unidentified growth factors—as mammals but the proportions differ.

Water

Water is the largest single constituent of animal tissue, amounting to approximately 58% in chickens and 66% in eggs. It is used for temperature regulation, it serves as a medium and transportation agent for dissolved substances, and it gives animals form and shape. It is the cheapest and one of the most important of all nutrients. Table 8.1 lists water consumption by chickens and turkeys of different ages under average conditions. These amounts vary depending on the food consumption, temperature, humidity, activity of the chicken, and nature of the food. One rule is that poultry will consume 1.75 units of water by weight for each unit weight of feed consumed at temperatures below 75°F (24°C) and 3.0 units of water per unit weight of feed at temperatures of 95°F (35°C).

Fresh, clean, uncontaminated water should be easily and readily accessible, both in quantity and location. For starting chicks, one fount

TABLE 8.1. Water Consumption[a] by Chickens and Turkeys of Different Ages (1000 Birds per Day)[b]

Age (weeks)	Chicken broilers (gal.)	Egg strain pullets (gal.)	Turkeys (gal.)
1	6[c]	5[c]	10[c]
2	12[c]	10[c]	20[c]
3	17	12	30
4	34	17	40
5	38	22	50
6	47	25	60
7	56	28	75
8	64	30	95
9		35	115
10		38	125
12		40	150
15		42	160
20		45	
35	Laying or breeding	50	Male 185 Female 119

[a] Will vary considerably depending on temperature.
[b] Source: Anonymous (1985).
[c] Consumption level at the end of the week.

per hundred chicks or an equivalent space depending on watering equipment used, evenly distributed and close to the heat sources, is considered a minimum. Cool, not cold or hot, water is the most desirable. Extremely cold or hot water decreases water consumption and may lead to dehydration and slow growth of the chicks.

Water quality is also important. Water should be tested to see that salts, pesticides, and microorganisms are at an acceptable level and that the water is palatable to poultry. Water that adversely affects growth, reproduction, or productivity cannot be used. The birds will not grow, reproduce, or stay healthy and in some cases they may absorb harmful chemical residues, for example, pesticides, in their tissues.

Nitrates and nitrites in some water supplies can be a problem. Poultry can often tolerate ingestion of up to 300 ppm of nitrates or 100 ppm of nitrites. To provide a reasonable margin of safety the water supply should contain less than 100 ppm of nitrates and 10 ppm of nitrites.

Sulfates at a level above 500 ppm have a laxative effect on poultry and above that level produce severe diarrhea. Special treatment is required to reduce sulfate effects because home water softeners will not remove sulfates.

Salt water toxicity can upset the electrolyte balance of poultry and cause dehydration. Although it takes levels of 5000 ppm to influence palatability, the physiological effects on the birds begin at about 2860 ppm.

Drinking water from wells, farm ponds, and other sources contaminated with bacteria not only can cause disease but also can influence growth and the general well-being of the flock at a subclinical level. Contamination can occur not only at the well or pond source but also in the drinking fountains if not kept clean and sanitary. Bacteria readily reduce nitrates to nitrites. Chlorination is the most practical way to eliminate contamination and oxidative changes. Water provided to poultry ideally should be of quality fit for human consumption.

Carbohydrates

Carbohydrates are organic chemical compounds structured from carbon, hydrogen, and oxygen. They are used as a source of energy by the organism because they are easily burned in the body to produce heat and energy for body movement and function. Sugars, starches, and fiber are the most common forms in which carbohydrates are found in feeds. Glycogen is a carbohydrate synthesized in the body and stored in small amounts in the liver and muscles. It can be utilized rapidly under emergency and stress situations in the bird. Excesses of carbohydrates are converted to fats.

The important forms of carbohydrates are those which can be digested by the enzyme systems in the bird. For this reason the term "nitrogen-free extract" is used to refer to the soluble and digestible portion of carbohydrates. The undigestible carbohydrates, which are the structural components of plants, are referred to as fiber. Fiber is also important as an aid to normal digestive functions and it may also affect absorption of certain minerals.

Fats

Fats also consist of carbon, hydrogen, and oxygen but in ratios that are different from carbohydrates. They contain about two and one-quarter times the energy value of carbohydrates. Fats can be formed from carbohydrates and stored in the body. Generally speaking, fats also include oils, the only difference being that fat is solid and oil is liquid at room temperature (65°F, 18°C). Fats are further classified as vegetable fat or oil and animal fats and oils.

Fats are composed of fatty acids and glycerol. Some have a saturated chemical structure and others are unsaturated. They are also classified as essential and nonessential fatty acids. The polyunsaturated linoleic and arachidonic acids are considered to be "essential fatty acids." They have specific functions in the body that are not related to energy production. Birds exhibit poor growth, fatty livers, reduced egg size, and poor hatchability without these essential fatty acids. Other types of fats contain phosphorus and are known as phospholipids.

The type and condition of the fats consumed by poultry will also have a marked influence on the type of fat produced in the carcass. For example, if the fats are soft, the carcass fat will be soft. Poultry consuming large amounts of fish oil or rancid fats can have fishy or rancid flavors or undesirable flavors.

Fat is considered to be all the material in the carcass that will dissolve in ether. It is referred to as crude fat. Small amounts of fat are generally added to rations in addition to that supplied by other feed ingredients. Processed fats added to rations may readily become rancid and form peroxides that destroy vitamins A and E. Antioxidants are usually added to delay or prevent rancidity.

Proteins

Proteins contain carbon, hydrogen, oxygen, nitrogen, and sometimes sulfur. They differ chemically from carbohydrates and fats in that all proteins contain nitrogen. About 16% of protein is composed of the nitrogen molecule. Because proteins are the principal constituents of organs and soft tissues of the body they are needed for proliferation of new tissue (growth) and replacement of old or damaged tissues (injury or disease).

Proteins are classified chemically according to their solubility in dilute salts and strong alkalies. From a commercial standpoint they are classified by the number of amino acids that can be used by poultry and by their digestibility, which is measured by the ability of the bird's enzyme system to break them down. For example, keratin, a form of protein found in hair, hooves, toenails, and feathers, requires acids or heat under pressure to be broken down into simpler proteins that can be assimilated by poultry. Despite the amino acid qualities of this group of proteins with processing, they provide no nutritive value because poultry enzyme systems cannot break them down. Proteins can also be part of other nutrient complexes, for example, lipoproteins. Hemoglobin, a protein present in red blood cells, is a complex iron protein moiety.

When proteins are digested or hydrolyzed by heat enzymes or chemicals, the basic units or amino acids are the end result. The quality or usefulness of a protein depends on the number and proportions of required amino acids present. Although a protein may show a high essential amino acid content the practical usefulness of the protein can be determined only by feeding it. For example, vegetable protein sources are generally low or deficient in sulfur-containing amino acids which are found to a greater extent in animal proteins.

Although 23 chemically different amino acids have been isolated, poultry require only 11 of them. The essential amino acids for poultry are phenylalanine, isoleucine, lysine, threonine, histidine, arginine, tryptophan, methionine, valine, leucine, and glycine. Glycine is required only by the growing chick. Proline, a twelfth amino acid, may be required under some conditions.

The amount of protein in a feed can be calculated by determining its nitrogen content and then multiplying by a factor of 6.25 to give an estimated crude protein content. Although a protein may show a high amino acid content, it may not be a high-quality protein for poultry. The practical usefulness of the protein can be determined only by feeding it, though chemical tests may give some indication. The result is called the biological value. Any protein in excess of the amounts needed for growth and tissue repair is burned and used to supply energy, the same as carbohydrates and fats. Because carbohydrates and fats are more economical than protein they are used more often to provide energy.

Minerals

Minerals, unlike carbohydrates, fats, and protein that are organic, are inorganic. They exist primarily in the form of salts and ash. They are essential for bone and eggshell formation, and for regulatory processes in the body. The essential minerals required by poultry include calcium, phosphorus, magnesium, manganese, sodium, potassium, iron, copper, zinc, sulfur, fluorine, chlorine, iodine, selenium, and molybdenum. Macrominerals such as calcium, phosphorus, and sodium are required in the largest amounts and must be included as specific additives in the ration. Frequently manganese is deficient and supplementation is necessary.

Calcium

Calcium is required for bone and eggshell formation and for clotting of blood and muscular actions. Lack of sufficient calcium causes rickets, which results in soft rubbery bones and thin-shelled eggs. Feed grade limestone and marine shells such as oyster and clam shells are common sources of calcium used in poultry rations which are in the form of calcium carbonate ($CaCO_3$).

Phosphorus

Phosphorus is also an important constituent of bone and occurs in a fixed ratio with calcium; this ratio is referred to as the calcium–phosphorus ratio. Calcium, phosphorus, and vitamin D are closely interrelated and interact so that it is essential that all three be present in the proper levels. The acidity of the digestive tract and other minerals present also influence the amounts of calcium and phosphorus required. Dicalcium phosphate, bone meal, and defluorinated rock phosphate are principal phosphorus supplements in poultry rations. Although plants contain phosphorus it is largely in the organic form, phytin, which is a poor source of phosphorus for birds because of limited utilization.

Sodium and Potassium

Both sodium and potassium are essential in maintaining an electrolyte balance in the body. They maintain tissue fluids close to a neutral pH position so that these fluids do not become too acid or too alkaline. Poultry deficient in sodium have poor growth, a nervous condition that frequently brings on cannibalism, and poor utilization of feed. Potassium deficiency is characterized by reduced growth and an increase in kidney urinary nitrogen excretion. Potassium is found in and available to poultry from feedstuffs of plant origin in sufficient quantities so that no supplemental feeding is required. Sodium is added in small amounts in the form of common salt, sodium chloride.

Magnesium

Magnesium is an important constituent of tissues and body fluids. It combines with phosphorus in bone tissue. Magnesium ions also serve as important activators of some enzymes. Since natural feedstuffs most often contain sufficient quantities of magnesium for poultry no supplements are generally required. In some cases when feeding dolomitic limestone, which has a high magnesium content, toxicity occurs. The magnesium interferes with normal calcium usage and acts as a laxative.

Sulfur

Sulfur is required to carry out several specific reactions in poultry metabolism such as the synthesis of taurine. Generally sulfur is added to rations in the form of calcium sulfate at levels of 1% or less, or can be obtained from the metabolism of cystine and methionine at a greater cost.

Trace of Microminerals

Several minerals are required in very small amounts, hence the name trace minerals. They consist of iodine, copper, manganese, iron, zinc, cobalt, and selenium. Table 8.2 lists the roles played by these minerals.

Of these minerals, manganese often is not found in sufficient amounts in feedstuffs to provide a balanced ration. A deficiency results

TABLE 8.2. Function of Trace Elements[a]

Mineral	Essential compounds or enzymes
Iodine	Constituent of thyroxine
Copper	Amino oxidase, tyrosinase, ascorbic acid, oxidase
Manganese	Arginase, peptidases, kinases, decarboxylase, deoxyribonuclease
Iron	Cytochrome enzymes, catalase, peroxidase, hemoglobin
Zinc	Carbonic anhydrase, peptidases, lactic acid dehydrogenase, glutamic dehydrogenase
Cobalt	Constituent of cyanocobalamin (vitamin B_{12})
Selenium	Free radical control (glutathione peroxidase, pancreatic lipase)

[a] Source: Patrick and Schaible (1980).

in a condition known as perosis or slipped tendon. Therefore, manganese is generally added to rations as manganese sulfate.

There is a range in which all of these minerals must be available to poultry. Below this level deficiencies occur; above the top level of the range toxicity symptoms begin to occur. For example, the selenium content of grains grown in certain soils with high selenium content can be toxic to poultry, but this level is 500–1000 times the requirement.

Vitamins

Vitamins are complex organic structures essential in minute amounts for the growth, reproduction, and overall health of poultry. Some vitamins are dietary essential and some are metabolic essential. For poultry, vitamins are all literally dietary essential. Their principal role is body regulation rather than the development of the structural components of the body. Table 8.3 shows the vitamins needed by poultry, their sources, and deficiency symptoms. Table 8.4 lists the amounts of vitamins, linoleic acid, and minerals required by poultry of various ages.

Energy

A source of energy is the largest single dietary need for poultry. The true energy value can be determined only by feeding it to poultry. Energy is needed by the bird for basal metabolism (cellular activity, respiration, and circulation), voluntary activity, digestion and absorption, thermal regulation, waste formation and excretion, formation of tissue and feathers, and reproduction.

The first use of any feed consumed is for body maintenance functions. When this need is met energy can be used for growth followed by use for reproduction and, finally, when all these previous needs are met energy is stored in the body as fat.

A number of terms have been created to describe the idealized flow of energy through an animal.

Gross Energy (GE)   Gross energy is the caloric value of the foodstuff consumed by the bird determined by burning the feed in a calorimeter. It is the first measurement in a nutrition evaluation.

Digestible Energy (DE)   Digestible energy is the intake of energy minus the fecal energy of food origin.

Metabolizable Energy (ME)   Metabolizable energy is the energy in the food less the energy lost in feces, urine, and combustible gas.

TABLE 8.3. Vitamins Needed by Poultry[a]

| Vitamin | Deficiency symptoms | Sources[b] | Type of poultry ration usually requiring supplementation | | | |
			Starting	Growing	Laying	Breeding
Vitamin A	Chicks—wobbly gait, deposits of urates in kidneys, pustules in mouth Hen—reduced egg production, nutritional roup	Green pasture, alfalfa meal, corn gluten meal, yellow corn, fish oils, pure vitamin concentrates	Yes	Yes	Yes	Yes
Vitamin D$_3$	Chicks—leg deformities, soft bones (rickets), reduced growth Hens—poor eggshell formation, reduced egg production and hatchability	Fish oils, activated animal sterols, and sunlight	Yes	Yes	Yes	Yes
Vitamin E	Chicks—poor coordination due to brain degeneration, accumulation of tissue fluids Hens—poor hatchability	Alfalfa meal, vegetable oils, wheat germ, and pure vitamin concentrates	Yes	No	No	Yes
Vitamin K	Chicks—hemorrhages due to failure of blood to clot Hens—same as for chicks except that condition is rarely seen	Green pasture, alfalfa meal, chemically pure vitamin substitutes	Yes	Yes	Yes	Yes
Thiamine (B$_1$)	Chicks—loss of appetite, head retractions, loss in body weight Hens—same as above and stop laying eggs	Grains and grain products, oil seed meals, milk products, and pure vitamin	Rarely	No	No	No
Riboflavin (B$_2$)	Chicks—curled toe paralysis and reduced growth Hens—poor hatchability with many embryos dying during second week of incubation	Alfalfa meal, milk products, distiller's solubles, fermentation products, and pure vitamin	Yes	Yes	Yes	Yes

(continued)

TABLE 8.3. (continued)

Vitamin	Deficiency symptoms	Sources[b]	Type of poultry ration usually requiring supplementation			
			Starting	Growing	Laying	Breeding
Pantothenic acid	Chicks—poor growth, ragged feather development, and degeneration of skin around beak, eyes, and vent Hens—reduced hatchability	Alfalfa meal, dried milk products, fermentation residues, and pure calcium salt of vitamin	Yes	Yes	Yes	Yes
Niacin	Chicks—retarded growth and inflammation of mouth and tongue Hens—no symptoms observed in hen except on protein-deficient diet	Wheat and wheat milling by-products, soybean meal, and pure vitamin	Yes	Yes	No	Yes
Pyridoxine (B$_6$)	Chicks—poor growth, lack of coordination and convulsions Hens—reduced body weight, egg production, and hatchability	Milk products, meat and fish by-products, soybean meal, and alfalfa meal	Sometimes	No	No	Yes
Choline	Chicks—retarded growth and "slipped tendon" Hens—no deficiency known	Fish products and pure vitamin	Yes	Yes	Sometimes	Sometimes
Biotin	Chicks—cracking and degeneration of skin on feet, around beak, and slipped tendon Hens—reduced hatchability	Grains, soybean meal, alfalfa meal	Yes	No	No	Yes
Folacin	Chicks—poor growth, poor feathering, and anemia Hens—reduced hatchability	Alfalfa meal, wheat, soybean meal, and liver preparations	Yes	No	No	Yes
Cyanocobalamin (B$_{12}$)	Chicks—reduced growth Hens—poor hatchability	Fish meal, fish solubles, meat scrap, liver preparations, and fermentation products	Yes	Yes	No	Yes

[a] Adapted from Naber and Latshaw (1979).
[b] Dried brewers' yeast is an excellent source of many of the water-soluble viatmins including thiamine, riboflavin, pantothenic acid, niacin, pyridoxine, - biotin, and folacin. Its use may be considered when feeds require supplementation with a large number of the water-soluble vitamins.

TABLE 8.4. Nutrient Requirements of Leghorn-Type Chickens as Percentages or as Milligrams or Units Per Kilogram of Diet

Energy base kcal ME/kg Diet[a] →		Growing			Laying	Laying Daily intake per hen (mg)[b]	Breeding
		0–6 Weeks 2,900	6–14 Weeks 2,900	14–20 Weeks 2,900	2,900		2,900
Protein	%	18	15	12	14.5	16,000	14.5
Arginine	%	1.00	0.83	0.67	0.68	750	0.68
Glycine and Serine	%	0.70	0.58	0.47	0.50	550	0.50
Histidine	%	0.26	0.22	0.17	0.16	180	0.16
Isoleucine	%	0.60	0.50	0.40	0.50	550	0.50
Leucine	%	1.00	0.83	0.67	0.73	800	0.73
Lysine	%	0.85	0.60	0.45	0.64	700	0.64
Methionine + cystine	%	0.60	0.50	0.40	0.55	600	0.55
Methionine	%	0.30	0.25	0.20	0.32	350	0.32
Phenylalanine + tyrosine	%	1.00	0.83	0.67	0.80	880	0.80
Phenylalanine	%	0.54	0.45	0.36	0.40	440	0.40
Threonine	%	0.68	0.57	0.37	0.45	500	0.45
Tryptophan	%	0.17	0.14	0.11	0.14	150	0.14
Valine	%	0.62	0.52	0.41	0.55	600	0.55
Linoleic acid	%	1.00	1.00	1.00	1.00	1,000	1.00
Calcium	%	0.80	0.70	0.60	3.40	3,750	3.40
Phosphorus, available	%	0.40	0.35	0.30	0.32	350	0.32
Potassium	%	0.40	0.30	0.25	0.15	165	0.15
Sodium	%	0.15	0.15	0.15	0.15	165	0.15
Chlorine	%	0.15	0.12	0.12	0.15	165	0.15
Magnesium	mg	600	500	400	500	55	500

(continued)

121

TABLE 8.4. (continued)

Energy base kcal ME/kg Diet[a] →		Growing			Laying		Breeding
		0–6 Weeks 2,900	6–14 Weeks 2,900	14–20 Weeks 2,900	2,900	Daily intake per hen (mg)[b]	2,900
Manganese	mg	60	30	30	30	3.30	60
Zinc	mg	40	35	35	50	5.50	65
Iron	mg	80	60	60	50	5.50	60
Copper	mg	8	6	6	6	0.88	8
Iodine	mg	0.35	0.35	0.35	0.30	0.03	0.30
Selenium	mg	0.15	0.10	0.10	0.10	0.01	0.10
Vitamin A	IU	1,500	1,500	1,500	4,000	440	4,000
Vitamin D	ICU	200	200	200	500	55	500
Vitamin E	IU	10	5	5	5	0.55	10
Vitamin K	mg	0.50	0.50	0.50	0.50	0.055	0.50
Riboflavin	mg	3.60	1.80	1.80	2.20	0.242	3.80
Pantothenic acid	mg	10.0	10.0	10.0	2.20	0.242	10.0
Niacin	mg	27.0	11.0	11.0	10.0	1.10	10.0
Vitamin B 12	mg	0.009	0.003	0.003	0.004	0.00044	0.004
Choline	mg	1,300	900	500	?	?	?
Biotin	mg	0.15	0.10	0.10	0.10	0.011	0.15
Folacin	mg	0.55	0.25	0.25	0.25	0.0275	0.35
Thiamin	mg	1.8	1.3	1.3	0.80	0.088	0.80
Pyridoxine	mg	3.0	3.0	3.0	3.0	0.33	4.50

Source: Anonymous (1985).
[a] These are typical dietary energy concentrations.
[b] Assumes an average daily intake of 110 g of feed/hen daily.

Net Energy (NE)   Net energy is the metabolizable energy less the energy lost as heat.

Energy Feedstuffs

Feed ingredients and other additives are characterized by the general nutrient source or function they serve in the ration. In general, they are categorized as energy feedstuffs, animal and vegetable protein supplements, mineral supplements, vitamin supplements, and non-nutritive additives.

Cereal grains not only make up the chief sources of energy but also the largest part of a poultry ration. It is estimated that corn makes up about one-third of all feed consumed by poultry. Yellow corn is preferred because of its vitamin A content and xanthophyll pigment which is absorbed and gives the skin and egg yolks a nice yellow pigment.

Although it is not universally used, wheat may be used in poultry rations, particularly in the northwestern and Canadian wheat-growing areas. Generally, since it is used for human food, the price is too high for poultry rations.

Grain sorghums or milo are another widely used source of grains, particularly in the south and southwest areas. Other grains used in rations in the United States are oats and barley, though they are less desirable because of the fiber content of the hull.

Another source of energy is fats, which are often economical and in adequate supply. They reduce the dusty texture and improve the general appearance of the feeds. Problems with the use of fat are improper mixing and rancidity, which can influence the flavor of the poultry when they are fed fat. Antioxidants minimize the latter.

Animal and Vegetable Protein Supplements

The amounts and availability of the essential amino acids required by poultry determine the value and usefulness of protein sources. Generally, several different protein sources give a more adequate amino acid balance than one alone. Animal proteins are generally considered of higher quality than vegetable proteins since they contain minerals and vitamins and are generally richer in the sulfur-containing amino acids. However, they may vary more in composition and often are more expensive than vegetable proteins.

Meat or meat and bone meal is dry rendered bone, muscle, and organ tissue. Generally, meat and meat and bone meals are low in the essential amino acids, methionine, tryptophan, and isoleucine. Meat meal and meat and bone meal in addition to protein have some value from the fat that these meats contain.

Milk products used in poultry feeds are dried buttermilk, skim milk, and whey. They are low in arginine, glycine, and cystine but high in riboflavin. Milk products are often too expensive compared to other available animal proteins for use in least cost diets.

Fish meals come from fish processing wastes and certain species of small fish such as menhaden caught particularly for use as livestock feed. Fish meals have high protein content and a good amino acid profile.

Other animal protein sources used are blood meal, hydrolyzed poultry feathers, and meal from waste poultry parts and organs. Blood meal is deficient in isoleucine, hydrolyzed poultry feathers in methionine, lysine, tryptophan, and histidine, and poultry by-product meal in methionine and tryptophan.

Oil seed meals from soybeans, cottonseed, and peanuts are the residues left after the oil has been extracted. Soybean meal is the most widely used oil seed. It is deficient in methionine. Cottonseed meal is not only low in lysine and methionine but it also contains a toxic substance called gossypol, which must be reduced before use in poultry rations. Peanut meals are also low in lysine and methionine. Other less common oil seed proteins are sesame, rape, safflower, and mustard seeds.

Corn gluten meal, low in arginine, lysine, and tryptophan and dehydrated alfalfa meal are other products used partially as protein sources.

Mineral Supplements

Because calcium, phosphorus, sodium chloride, manganese, and iodine are not found in high enough amounts in normal feedstuffs, they must be added as supplements. Calcium is generally added by feeding ground limestone in the ration or crushed oyster or clam shells. Calcite, chalk, and marble are also used as calcium supplements.

Phosphorus is generally supplied as dicalcium phosphate or in the form of rock phosphates. Generally, rock phosphate must be defluorinated by heating.

Generally common salt needs to be added in amounts up to 1.5%, preferably in iodized form to also supply iodine.

Some poultry feeds contain other minerals such as copper, iron, selenium, and zinc and are usually added as a purchased mineral mixture.

Vitamin Supplements

Modern feeds are supplemented with vitamin mixtures that may be made from chemically pure sources of the vitamin.

Vitamin A is added as the pure vitamin or as fish oils. Vitamin D is usually added as irradiated fish oils or animal products, which give these products vitamin D activity. Poultry require a special vitamin D isomer that is found only in animal products. Vitamin D from plant sources is not satisfactory and cannot be efficiently utilized by poultry. Generally, riboflavin, niacin, pantothenic acid, choline, and vitamin B_{12} are added to the ration in the form of a vitamin premix.

Other Additives

Modern poultry diets may also include several nonnutritive feed additives.

Antibiotics and arsenicals are often added in parts per million as growth promoters. Antibiotics commonly used are bacitracin, lincomycin, oleandomycin, oxytetracycline, penicillin, and tylosin.

Drugs in the form of coccidiostats are routinely added to growing rations to prevent coccidiosis. Turkey rations frequently contain antihistamonad drugs to prevent blackhead.

Antioxidants are added to feeds to prevent fats and oils from becoming rancid and to prevent oxidation of several of the vitamins in the ration, which causes them to lose their vitamin activity.

Grit in the form of crushed granite is sometimes added as a supplement or component of poultry rations for the purpose of aiding the gizzard in grinding food materials.

Other nonnutritive feed additives include pigmentors for broiler rations, pellet binders, chelating agents, mold inhibitors, and organisms such as lactobacillus to name but a few that are available.

REFERENCES

Anonymous. 1985. Nutrient Requirements of Poultry, 8th Edition. National Academy of Sciences, Washington, DC.

Ensminger, M. E. 1971. Poultry Science. Interstate Publishing Co., Danville, IL.

Naber, E. C., and Latshaw, J. D. 1979. Poultry Nutrition and Feeding. Ohio State University Extension Bull. 602, Columbus, OH.

Patrick, H., and Schaible, P. J. 1980. Poultry Feeds and Nutrition, 2nd Edition. AVI Publishing Company, Westport, CT.

9

Diseases and Parasites of Poultry

The art and science of poultry disease control is as complex, variable, and confounded with as many apparently unrelated events as is the practice of human medicine. As more birds are grown in more concentrated areas and in tighter confinement, new disease problems appear and old ones sometimes reoccur. Fortunately, for the average poultryman, good management, the ability to detect disease or parasite problems at an early stage, and the knowledge and judgment to know when and where to go for help when needed should make it possible for him or her to cope successfully with most disease and parasite problems. In this chapter an attempt will be made to present the causes of disease and the basic concepts of disease prevention and control along with examples of the most serious and prevalent poultry diseases.

Disease is considered to be any deviation from a normal state of health. It can be caused by trauma or injury, nutrient deficiencies, microorganisms, such as protozoa, bacteria, mycoplasma, viruses, yeasts, and molds, internal and external parasites, behavioral and reproductive problems, and poisons. Almost all disease problems on poultry farms start from new poultry brought on the premises, contaminated premises from previous flocks, or lack of proper sanitation and other good management practices.

DISEASE PREVENTION

Schwartz (1977) has listed 15 fundamental factors of disease prevention in poultry health management.

1. Preventive medicine is the only effective approach in health management in today's intensive poultry operations. As flock size doubles, the possibilities of disease quadruple.
2. Preventive medicine includes disease-free chicks, maximum hygiene and care, vaccination, medication, and adequate nutrition—performed under conditions of strict sanitation and isolation (Fig. 9.1).
3. Change the litter and thoroughly clean and disinfect the house and equipment after each group of birds or when feasible.
4. Select healthy, vigorous, disease-free chicks, poults, or pullets.
5. Keep young birds well isolated from older birds. Separate caretakers and equipment will increase the chances of success in isolation. Maintain breeder flocks on separate premises.
6. Isolate poultry classes from other livestock. Chickens, cattle, turkeys, and swine are subject to cross-infections.
7. Provide adequate commercial feed or carefully formulated home-mixed feeds.
8. Provide a continuous supply of potable water for birds of all ages. In summer keep the water cool by protecting it from the sun. Protect it from freezing in winter. Birds consume up to two and one-half times as much water as feed. When water intake decreases, there is a proportionate decrease in feed intake too. A marked reduction in feed and water consumption is usually the first sign of illness.
9. Do not crowd poultry. Crowding increases cannibalism, feather picking, hysteria, and other stress-related problems. Crowding retards growth, reduces feed efficiency, and decreases production.
10. Have a *sound vaccination program* and follow it carefully. For young birds, raise house temperature 5°F during the vaccination reaction period. Schedule all vaccinations and revaccinations on farms with multiage flocks for the same day. Live-virus vaccines can spread to susceptible poultry.
11. During brooding, regulate temperature, humidity, and ventilation to the comfort of the chicks or poults. Prevent drafts, overheating, and chilling.
12. Keep unauthorized personnel out of the poultry house.
13. Incineration is the most satisfactory method of dead-bird disposal. Disposal pit and deep burying, in that order, are the next

FIG. 9.1. Cleanliness and sanitation are fundamental factors in disease prevention. (Source: USDA.)

best methods. Dead birds, if not properly disposed of, become a disease threat to *all* poultry in the area.

14. In disease outbreaks promptly obtain a reliable diagnosis. Then use the best treatment for control of that particular disease. Birds recovered from diseases such as pullorum and mycoplasmosis should be sold for slaughter—not saved for breeder replacements.

15. It is best to accompany diseased or dead birds to the diagnostic laboratory or be available by telephone; you will therefore be available to give additional information that might be needed.

DIAGNOSTIC ORGANS IN POSTMORTEM EXAMINATIONS

Any large poultry operation loses some birds. Postmortem examinations help to determine whether there is a problem and often offer tangible evidence of what type of disease is in the flock (Fig. 9.2). For that reason, it is important for a poultryman to know and become experienced in conducting postmortem examinations. Such examinations are in addition to but not in place of those done by a qualified

FIG. 9.2.   Performing a postmortem on a bird. (Source: USDA.)

veterinarian. The following procedure is recommended by Salsbury
Laboratories:

1. Cut the skin on the side of the mouth. Scan the mouth for lesions
   of pox, mycosis, and a number of other diseases.
2. Open the esophagus. Inspect it for injury from foreign matter on
   tiny nodules (bumps).
3. Slit the larynx and trachea lengthwise. Examine them for excess mucus, blood, and cheesy material.
4. Cut into the crop. Note whether the contents are impacted or
   sour smelling. Wash out the crop and examine the lining for
   signs of mycosis, etc.
5. Observe the condition of the air sacs, which are often cloudy in
   disease.
6. Examine the heart, liver, spleen, lungs, and bronchial tubes for
   obvious lesions.
7. Examine the intestines for nodules, tumors, or hemorrhages.

Slit the intestines lengthwise to examine for worms, free blood, inflammation of the lining, ulcers, hemorrhagic areas, and excess mucus.

8. Cut into the ceca, noting the nature of the contents. If blood is found, wash it out and examine the lining. Look for cheesy sores, tiny cecal worms, or a scarred lining.

9. Slit the proventriculus, noting any hemorrhages in the lining, or white coating.

10. Open the gizzard, examining the lining for roughness and erosion.

11. Look for abnormalities of the ovary, oviduct, and kidneys.

12. Look for a swollen brachial nerve, one of the lesions of leukosis.

IMMUNOLOGY OF DISEASE CONTROL

Birds, like other animals, have many mechanisms that provide natural disease resistance. These innate barriers include items such as the skin, cilia of the cells that line the upper respiratory tract, and other features of the body that either prevent the entry of foreign microorganisms or nonspecifically prevent their growth inside the body. The skin and cilia act as mechanical barriers to keep bacteria and other microorganisms out of the body. Because the body temperature of birds is often in excess of 40°C, bacteria that grow well at 37°C may not effectively reproduce inside birds, and so do not pose a serious disease threat. The effectiveness of these mechanisms can be compromised by poor nutrition and management practices.

Even with the above-mentioned natural barriers to infection, birds must possess an active immune mechanism or system to survive constant exposure to the vast numbers of disease-causing microorganisms and agents. Eventually bacteria, viruses, molds, fungi, and toxins gain entry to the body. The immune system, which must react to and neutralize these invaders, is made up of the white cells present in the blood and certain organs of the bird's body (e.g., spleen, thymus, bursa of Fabricius, cecal tonsils, and other lymphoid tissues).

Birds have several types of white cells (or leukocytes). Those in the blood are lymphocytes, monocytes, heterophils, basophils, eosinophils, and thrombocytes. These cells perform many functions that collectively help to prevent or cure diseases. For example, monocytes are very important since they can ingest (i.e., phagocytize) foreign particles and digest them. In addition, it is believed that monocytes can grow and become macrophages, i.e., very large cells capable of phagocytizing large quantities of bacteria and other particles. These bacteria are then digested by enzymes present in the macrophages and are thus removed from the body. Basophils are often involved in inflammatory immune responses as are monocytes, lymphocytes, and macrophages. There are at least two distinct types of lymphocytes in the body. One type originates in the bursa of Fabricius of young birds. These cells, called B cells, migrate to the spleen and other lymphoid organs of the birds and are capable of responding to bacteria, toxins, and other pathogens or foreign agents by producing proteins called antibodies. These antibodies will react with the particular substance that initially stimulated the B cells.

The second type of lymphocytes, the T cells, arise in the thymus of young birds. These cells also migrate to the spleen and other lymphoid tissues where they can respond. They are industinguishable in appearance from B cells, but perform different functions. The T cells do not produce antibodies; instead they produce proteins called lymphokines. Lymphokines (1) react with leukocytes and macrophages in the bird to increase their phagocytic removal of organisms, (2) cause lymphocytes to be produced rapidly, (3) nonselectively lyse or break down cells, (4) prevent viral replication, and (5) perform other related functions. In addition, B and T cells work together to cause removal of foreign cells and substances from the body. Neither type of lymphocyte alone is capable of preventing disease. T cells are more efficient in preventing viral diseases while B cells are more effective against bacteria.

To maintain flocks of chickens or other poultry in a relatively disease-free state, producers should promote the integrity of natural barriers to prevent microbial entry into the birds. In addition, it is important that sound vaccination programs be practiced, so that the B and T cell populations are continually stimulated to produce the antibodies and lymphokines that will fight and cause removal of disease-causing organisms that enter the body.

POULTRY DISEASE CONTROL STRATEGY

Avian medicine is concerned with the entire population of poultry more than it is with the individual bird. Disease prevention and control are more important in maintaining flock health than therapy and treatment of the disease. Normal disease controls are the husbandry practices utilized by the individual or firm that address preventing disease.

VACCINATION

Vaccines act as insurance against the risk of disease, but like insurance, they have a cost associated with them. These costs include the price of vaccine, time involved in scheduling and administering vaccines, as well as losses or reactions associated with the vaccine itself. Vaccines are normally used to prevent or reduce problems that can occur if a flock is exposed to the field strain of a particular disease.

Types of Vaccines

Live vaccines contain viruses or bacteria that infect the bird and multiply in its body to produce immunity. Most vaccines used today are modified or attenuated (artificially changed to reduce virulence) or avirulent (non-disease-causing). Killed vaccines (bacterins) are prepared from viruses or bacteria that have been inactivated by processing with heat, formalin, or β-propiolactone treatment. These vaccines are then usually combined with an adjuvant such as aluminum hydroxide or an oil to increase the stability of the killed antigen and to stimulate the immune system for a longer period.

Toxoids are vaccines prepared to stimulate immunity against a toxin, such as botulism. Toxins are treated to destroy their ability to cause disease but left recognizable as an antigen so that antibodies will be produced.

Methods of Vaccination

A variety of methods of vaccinating poultry have evolved depending on the vaccine utilized. These include vaccinating by subcutaneous or intramuscular injection, water vaccination, wing-web vaccination, and eye drop or spray vaccination.

Vaccination Program

Many factors are involved in determining when and if to vaccinate, what to use, as well as the route of administration. Some considerations are the types of diseases in the area, availability and cost of vaccines, climatic conditions, and management.

An example of a broiler vaccination schedule is given in Table 9.1.

NUTRITIONAL DEFICIENCY DISEASES

Nutritional deficiencies are rare in modern poultry operations. Exceptions occur when a ration is misformulated by mistake or accident and when other diseases or stress prevent normal consumption or absorption of feed. Table 9.2 lists the essential minerals and vitamins and their deficiency symptoms.

TABLE 9.1. Example of a Broiler Vaccination Schedule for Individual Bird Protection [a, b]

Age (days)	Vaccine	Method of vaccination
1	Marek's disease	Subcutaneous
	Newcastle disease B$_1$ strain	Spray
	Infectious bronchitis Mass. strain	Spray
	Infectious bronchitis Conn. strain	Spray
14–18 [a]	Newcastle disease B$_1$	Spray or water
	Infectious bronchitis Mass. strain	Spray or water
	Infectious bronchitis Conn. strain	Spray or water
	Infectious bursal disease—mild	Water

[a] Source: Harris (1967).
[b] Fourteen to eighteen day vaccinations in broilers vary depending on the disease prevalent in the area. Programs vary in the strain of infectious bronchitis used, the route of administration, and whether infectious bursal disease vaccine is used in the area.

PROTOZOAN DISEASES

Protozoa are microscopic single-celled animals. They cause two of the most common and oldest recognized diseases in poultry, coccidiosis in chickens and histomoniasis (blackhead) in turkeys. Less frequent protozoal diseases include trichomoniasis and hexamitiasis.

Coccidiosis infections in chickens are worldwide and are caused by nine species of the genus *Eimeria*, six of which are especially virulent. Seven other species infect turkeys, but all the organisms are host specific, that is, chickens do not become infected with the turkey species and vice versa.

Coccidiosis is spread primarily by birds eating droppings and contaminated litter. Birds with no access to litter and droppings do not become infected. This is one of the reasons why commercial egg type pullets and layers are kept in cages.

Young birds are most susceptible to coccidiosis. The symptoms are weakness, pale skin, ruffled feathers, chirping, and usually bloody droppings. Infected birds droop their wings, huddle together, and have little appetite for feed or desire for water. On autopsy, the intestine and ceca are distended and contain a foamy foul-smelling material. In advanced stages, bloody intestinal contents and pinpoint lesions occur on the walls of the intestine.

Prevention is the best method of handling coccidiosis. To avoid infection, commercial pullets and layers are managed to prevent access to litter and droppings.

For broilers, prevention also consists of feeding a coccidiostat at a continuous low level in the feed. Since broilers are reared on litter with access to the coccidia oocysts the management plan requires the use of a coccidiostat. Treatment of coccidiosis is costly and recovery of meat-type birds lengthens the growing period adding to the cost.

TABLE 9.2. Essential Minerals and Vitamins and Their Deficiency Symptoms[a]

	Functions	Deficiency symptoms
Minerals		
Macrominerals		
Calcium	In blood clotting, skeletal bone formation, strengthening egg shell	Rickets; young—osteomalacia; old—poor egg shell quality and hatchability
Phosphorus	In metabolism as high energy bone, bone formation	Rickets, poor egg shell quality and hatchability
Magnesium	In metabolism of carbohydrates and protein	Sudden, convulsive death
Microminerals		
Manganese	Enzymatic function	Perosis, poor hatchability
Iron	In cellular respiration	Anemia
Copper	In iron absorption, enzymatic function	Anemia
Iodine	As thyroid hormone	Goiter
Zinc	Enzymatic function	Poor feathering, short bones
Cobalt	As vitamin B_{12}	Slow growth, decreased feed efficiency, mortality, reduced hatchability
Molybdenum	Enzymatic function	—
Selenium	—	Exudative diathesis
Vitamins		
Fat soluble		
Vitamin A	Aids growth, vision, maintains epithelial tissue	Decreased egg production; xerophthalmia; ataxis, weakness; lack of growth
Vitamin D_3	Aids calcium and phosphorus absorption; bone formation	Thin-shelled eggs; reduced egg production and hatchability; retarded growth; rickets—young animals; osteomalacia

	Function	Deficiency symptoms
Vitamin E	Biological antioxidant, to help maintain reproductive capability	Enlarged hocks; encephalomalacia (crazy chick disease)
Vitamin K	In blood coagulation; may be involved in oxidative respiration	Prolonged blood clotting, intramuscular bleeding
Water soluble		
Thiamine (B_1)	In carbohydrate and fat metabolism	Loss of appetite; polyneuritis and death
Riboflavin (B_2)	In energy metabolism	Curled-toe paralysis; poor growth, egg production, and hatchability; dermatitis
Pantothenic acid	In protein, fat, and carbohydrate metabolism	Mild dermatitis; crusty scab-like lesions at corners of mouth and on feet
Niacin or nicotinic acid	In protein, carbohydrate, and fat metabolism	Enlarged hocks; bowed legs, diarrhea, inflammation of tongue and mouth cavity
Pyridoxine (B_6)	In protein metabolism	Reduced egg production and hatchability
Choline	In nerve impulse transfer	Poor growth, fatty liver, decreased egg production, perosis
Vitamin B_{12}	In red cell formation, carbohydrate and fat metabolism	Pernicious anemia, poor growth, embryonic mortality
Folic acid	In red cell formation, protein metabolism	Poor growth, anemia, poor feathering, egg production, and hatchability; cervical paralysis in poults
Biotin	As antidermatitis factor	Dermatitis on feed and around beak and eyes; perosis
Vitamin C	May aid in egg formation during heat stress	Not demonstrated
Inositol	In fat metabolism	Poor growth; fatty liver

[a] Source: Patrick and Schaible (1980).

Blackhead or histomoniasis is primarily an intestinal disease mostly of young turkeys. It is widespread in the United States. The disease is caused by the organism *Histomonas meleagridis*. Blackhead is spread by a parasite of turkeys, cecal worms *(Heterakios gallinarum)*. The protozoan organisms, in combination with the bacterium *Escherichia coli,* which is part of the usual gut microflora, cause characteristic "blackhead" lesions to develop.

When these protozoan organisms are passed in the droppings of infected birds and in the eggs of the cecal worm, other birds pick them up from the droppings and litter. Since earthworms feed on cecal worm eggs, they perpetuate the disease and also become a source of infection. Although blackhead can be a particular problem with range turkeys it also affects pheasants and peafowl.

Blackhead gets it common name from the fact that the heads of dead birds turn dark purple or black, a condition called cyanosis. Infected birds have an increased desire for water, decreased appetites, and are drowsy, weak, and droop their wings. Droppings are brownish yellow, watery, and foamy. In young birds mortality frequently reaches 50%.

On postmortem examination, typical caseous or cheesy lesions are found in the lower intestines, ceca, and liver. Liver lesions are also yellowish and circular with depressed centers. These liver lesions look almost like daisies in severely damaged livers and are often called "flowers of blackhead." The intestinal contents contain a caseous like core.

Treatment consists of blackhead drugs or histomonostats mixed in the feed or water for about 5–7 days. Histomonostats can also be used continuously in feeds for turkeys over 6 weeks old to prevent outbreaks. Blackhead, though a serious disease problem, can be avoided or controlled with a sound preventive health plan.

Trichomoniasis is a protozoan disease caused by *Trichomonas gallinae.* It is found in turkeys, chickens, and pigeons. It can be identified by raised caseous lesions in the crop, mouth, and esophagus. The disease is often found in small flocks of poultry and outbreaks are usually associated with unsanitary conditions. Prevention by good management is the best policy to follow. Treatment by copper sulfate at 1:2000 dilution in the water or at 2 lb/ton of feed for 3–4 days is usually recommended. Dimetridazole is also reported to be an effective control.

Hexamitiasis is an acute infectious disease caused by *Hexamita meleagris.* It is found primarily in turkeys under 10 weeks of age but it can also affect quail, ducks, pigeons, peafowl, and partridges. Affected birds show subnormal temperature and frequently chirp. Yellow-colored, watery diarrhea is associated with the disease. Foamy watery mucus is also found in the duodenum and upper intestine. Furazolidone and oxytetracycline are frequently used treatments.

BACTERIAL DISEASES

Bacterial diseases that affect poultry are numerous. One method of classification is to consider the system the organism affects. Examples of respiratory bacterial diseases include colibacillosis, infectious coryza, and fowl cholera infections. Examples of bacterial infections affecting internal organs are paratyphoid, pullorum, fowl typhoid, and omphalitis.

Colibacillosis is characterized by any one of a group of infectious diseases in which *E. coli* is the primary or secondary causative agent. *E. coli* is part of the natural gut flora and is an opportunistic pathogen that can become a problem when stress or disease occurs. All birds are carriers. Diagnosis is through laboratory isolation of the coliform organism. Many antibiotics and drugs are used for treatment but preventive management such as good sanitation and minimizing stress are still the best procedures.

Infectious coryza is a respiratory disease caused by *Haemophilus gallinarum,* a gram-negative nonmotile bacteria. The disease is characterized by nasal discharge, swelling of the face, and sneezing. Lesions include inflamed nasal passages and sinuses with discharges of yellow mucus and cheesy exudates in the cavities. Treatments are numerous and vaccination is available. All-in, all-out rearing and keeping young bird flocks away from older flocks is one control method.

FOWL CHOLERA

Fowl cholera is caused by the bacterium *Pasteurella multocida,* a gram-negative bipolar staining bacillus. The disease was observed as early as 1736. The organism infects all species of poultry worldwide and is becoming an increasing problem. Birds approaching maturity or adults are most often affected. The organism is very susceptible to common disinfectants.

Transmission is by other poultry, wild birds, predators, and rodents. Infection can occur through the respiratory tract, eyes, or open wounds.

One characteristic feature of fowl cholera is that it occurs rapidly. Birds will be found dead with no explanation. As the disease progresses, birds lose weight, decrease feed and increase water consumption, have pale yellow droppings, and sometimes produce a rattling noise from mucus in their air passages. A differential diagnosis is necessary because symptoms can be similar to other respiratory diseases. Positive identification is made by the presence of bipolar staining organisms in the blood and isolation of the causative agent. Flocks can be immunized with a three-strain cholera killed bacterin or a live "Cu" vaccine. Treatment products include sulfadimethoxine, sulfaquinoxaline, and oxytetracycline.

NONRESPIRATORY BACTERIAL DISEASES

Paratyphoid infections affect many birds and mammals and are caused by several groups of salmonellae that are not host specific. The pathogenic effect of paratyphoid organisms stems from the endotoxins they produce. These species of salmonellae are motile nonhost adapted and number between 10 and 20 species. Salmonellae tend to be susceptible to disinfectants and formaldehyde gas. Contaminated feed ingredients are one source of infection. Feces, infected feed, water, or contaminated litter are other sources. The salmonellae localize in the gut and are eliminated intermittently by the host, which becomes a carrier. Paratyphoid is generally a disease of young birds, which show profuse diarrhea, dehydration, shivering, and huddling near the heat source. Posted birds show severe enteritis, and swollen livers, spleens, and kidneys. Treatment is designed to hold the disease in check to permit marketing of recovered birds. Many drugs and antibiotics have been used with some success.

Pullorum disease, often called white diarrhea, is primarily an egg-transmitted disease caused by *Salmonella pullorum*, a nonmotile, gram-negative bacillus. Young chickens are highly susceptible as well as other avian species and some mammals. Vertical transmission from hen to egg to chick is the most important route. Horizontal spread in the hatchery from contaminated incubators, hatchers, and chick boxes is another way. Mortality in young birds is high around 4 or 5 days of age. Chicks show pasted vents, chalk-white feces stained with bile, and increased water consumption.

Prevention is based on establishing and maintaining pullorum-free breeder and multiplier flocks. Stained-antigen, rapid, whole blood testing is used for chickens (Fig. 9.3). Tube agglutination tests utilizing sera are used for turkeys and for confirming plate test reactions. Detailed regulations for control of pullorum are given in the National Poultry Improvement Plan, ARS, BARC-East, Bldg. 265, Beltsville, MD 20705. Treatment has been helpful in lowering mortality by the use of furazolidone, sulfa drugs, or broad-spectrum antibiotics. Water medication is preferred to feed medication because infected birds stop eating.

Pullorum can be transmitted to humans by uncooked eggs and poultry meat. In humans it causes acute gastroenteritis or upset stomach. Recovery usually occurs in a few days.

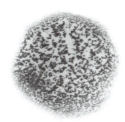

FIG. 9.3. Reactions to the stained-antigen, rapid, whole-blood test for pullorum disease. Left, negative reaction; center, questionable reaction; right, positive reaction. (Source: USDA.)

Salmonella gallinarum is the causative bacterium of the disease called fowl typhoid. It is a septicemic infectious disease mainly of chickens and turkeys. Since the antigenic reactions in birds are the same as for pullorum, reactors are identified during pullorum testing.

Unlike pullorum, the age of highest susceptibility is 2 weeks or older with mortality rates of 5–30%.

Transmission is the same as pullorum. It is not possible to distinguish pullorum from typhoid by visual symptoms. Positive diagnosis is by actual isolation and identification of the *S. gallinarum* organism.

Acute fowl typhoid lesions include bile-stained enlarged livers, enlarged spleen and kidneys, and pallor throughout the body. Prevention and treatment are the same as for pullorum. Pullorum and typhoid are mandatory reportable diseases to the State Veterinarian under regulations established under the National Poultry Improvement Plan. Killed bacterin vaccines are no longer available under United States licensing laws.

Omphalitis or navel ill is an inflammation of the navel involving bacterial infection caused by improper closure of the navel. The bacteria involved may include *E. coli, Pseudomonas, Salmonella,* or *Proteus.* Chicks hatched from dirty eggs or near exploded eggs are likely candidates for omphalitis. The disease is related to poor incubator and hatchery sanitation, excessive incubator humidity, chilling, and overheating. Poorly managed breeder flocks that allow fecal contamination of eggs is also a source. Mortality may reach 10–15% but can be lessened by using a broad-spectrum antibiotic in the drinking water.

MYCOPLASMA

Mycoplasmosis is a serious contagious disease that affects poultry at any age. Some 20 different serotypes of mycoplasma have been isolated from avian species. The following three types are pathogenic: *Mycoplasma gallisepticum, Mycoplasma synoviae,* and *Mycoplasma meleagridis.* Mycoplasma is a scientific generic name for a group of organisms that are smaller than bacteria but larger than viruses. Mycoplasma are species specific and they infect humans, livestock, and poultry. Because mycoplasma have thin cell walls, they are easily killed by disinfectants when outside the host.

Mycoplasma gallisepticum (MG) was called chronic respiratory disease before the causative agent was found. The disease causes respiratory distress and air sac lesions in birds. Turkeys may show signs of sinusitis. The organism may be present in the bird and show no signs of disease until the bird undergoes stress. Airsacculitis is the term used to describe MG infections complicated by a respiratory virus. *E. coli* is an opportunistic bacteria that is often found along with mycoplasma. Prevention is initiated by blood testing breeders and using only MG clean breeding stock. Infected flocks should be marketed as soon as feasible. Treatments effectively used include tylosine, oxytetracycline, and gallimycin.

Mycoplasma synoviae (MS) or infectious synovitis is characterized by inflammation of the synovial membranes of the joints and tendon sheaths. MS affects layers, breeders, broilers, turkeys, and even pheasants and geese. The spread of the disease is usually by direct contact from the respiratory tract or indirectly from such things as artificial insemination of turkeys with semen from affected toms. Egg transmission from infected breeder flocks is also possible.

Prevention includes blood testing of breeders and using only clean breeder flocks. Good sanitation and isolation should reduce the odds of introducing the disease. Chlortetracycline is one drug used for treatment.

Mycoplasma meleagridis (MM) is an organism that is host specific for turkeys. MM is an egg-transmitted and venereally transmitted disease. The disease may end as airsacculitis. Poor performance in turkeys and increased skeletal abnormalities are signs of MM. Egg transmission is probably the primary source of infection. Turkey hens commonly become infected when inseminated with semen from infected toms. Prevention should include blood testing and care and screening of breeder toms to keep from infecting the hens during insemination. Treatment is similar to that for MG and can include tylosine, oxytetracycline, or gallimycin.

VIRAL DISEASES

There are 17 or more viral diseases that can affect poultry. Examples are Newcastle, laryngotracheitis, infectious bronchitis, fowl pox, infectious bursal disease, avian influenza, Marek's disease, and lymphoid leukosis (Fig. 9.4).

Newcastle disease, so called because it was first isolated at Newcastle upon Tyne in England, is a contagious disease of the respiratory tract with pronounced nervous symptoms. Newcastle affects all birds, but chickens and turkeys are the most susceptible poultry. Man and other mammalian species can be infected by the virus. The etiologic agent is an RNA paramyxovirus existing in more than 100 strains. The lentogenic (mildly pathogenic) and mesogenic (moderately pathogenic with respiratory distress) forms of Newcastle are most com-

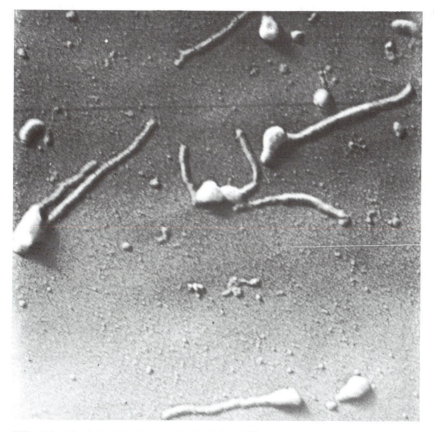

FIG. 9.4. Electron microscopic photograph of Newcastle virus particles. ×122,000. (Source: USDA)

velogenic (highly pathogenic with high mortality) form of Newcastle is the most feared in the poultry industry. Prevention is by vaccination; however, vaccines do not protect fully against the velogenic (highly pathogenic) form.

Infectious laryngotracheitis or avian diphtheria is a highly contagious disease which primarily affects mature chickens. The symptoms are severe respiratory distress and bloody tracheal exudates. The disease is caused by a "group A" herpesvirus (DNA). The virus is easily destroyed by disinfectants and ultraviolet light but can survive for a long time in dead birds or tracheal expectorants. Chickens, pheasants, and peafowl are susceptible. The trachea contains bloody mucus in affected birds as well as cheesy plugs. Death is usually caused by asphyxiation. Vaccination with an attenuated vaccine is the conventional form of prevention in problem areas. No specific treatment of laryngotracheitis is available.

Infectious bronchitis is a highly contagious disease manifesting itself by respiratory signs, gasping, and coughing. Turkeys are naturally resistant but chickens of all ages are affected. Infectious bronchitis is caused by an RNA-based coronavirus that has many serotypes. Mortality can be high in chicks. Transmission can be direct from bird to bird or through common fomites such as vehicles and egg cases. Vaccination with a modified live virus that contains the serotype found in that area is the best prevention. No specific treatment is used, except that broad-spectrum antibiotics are administered to prevent complications from secondary infections.

Fowl pox is a chronic contagious disease that occurs in a diphtheric (wet) form and a cutaneous (dry) form that can affect most species of birds of all ages. The disease can be transmitted directly from bird to bird. The virus itself usually invades the body through skin abrasions. Mosquitoes of the *Culex* and *Aedes* genera are mechanical carriers. Mosquitoes are often associated with pox problems of range turkeys in the fall of the year. Wet pox show signs of raised whitish-yellow plaques visible in the mouth. Dry pox often result in yellow-brown wart-like masses on the unfeathered portions of the head, comb, or face. No known treatment exists but vaccination during an outbreak helps to reduce losses. Vaccination is accomplished by using fowl pox and pigeon pox vaccines as a primary method of prevention.

Infectious bursal disease is a contagious disease infecting the bursa of Fabricius of chickens. It tends to occur in 3- to 6-week-old chickens and is thought to be a diplornavirus. The disease appears to affect egg-type chickens more than meat-type birds. Transmission can be direct from bird to bird or indirect through feed and water contaminated with feces. Darkling beetles are proven vectors. Lesser meal worms also harbor the virus. Posted birds show signs of bursal atrophy after having been enlarged. The kidneys and ureters are enlarged and distended with urates. There is no treatment but good management can help reduce the severity of the disease. Vaccination of the breeder can result in chick protection for up to 2 weeks through the maternal antibodies. Presently, some states do not allow infectious bursal disease vaccination.

Avian influenza, often called fowl plague in the past, is a contagious disease that infects chickens, turkeys, ducks, and wild birds. It is a type A influenza virus that has several serotypes. It can be spread from bird to bird through fecal or airborne transmission as well as by clothing, vehicles, or other fomites. Recovered birds shed the virus for several months. Migratory waterfowl and sea birds can carry the disease but show no clinical signs. Positive identification requires viral isolation. Quarantine, depopulation, and clean-out of houses are used when highly pathogenic strains are encountered. Vaccines are generally not useful because of the antigenic variety within strains. Vaccination complicates the problem further because vaccinated birds test positive for the disease. All suspected outbreaks should be reported to the state veterinarian.

Marek's disease is a contagious disease of young chickens. It is a neoplastic disease caused by a cell-associated DNA herpesvirus that is antigenically related to the turkey herpesvirus. The primary route of transmission is through inhalation of dander from the feather follicles of infected birds. Droppings and litter can remain infective for up to 16 weeks. Birds can shed the virus from the skin for up to 76 weeks. The darkling beetle is also suspected as a vector. Upon posting, several types of lesions may be apparent (Fig. 9.5), peripheral nerves are enlarged and discolored, and lymphoid tumors occur in internal organs in the visceral form (Fig. 9.6). No treatment exists but prevention can be accomplished by vaccinating day-old chicks with an HVT (herpesvirus turkey origin) Marek's vaccine administered by subcutaneous injection on the back of the neck.

Lymphoid leukosis is a contagious neoplastic disease that generally affects chickens that are over 16 weeks of age. Tumors develop in the viscera, especially the liver and bursa of Fabricius. Lymphoid leukosis is caused by oncovirus C which is an RNA virus of the retroviridal family. Females appear to be more susceptible to lymphoid leukosis than males. The disease is characterized by rather persistent low mortality. Lymphoid leukosis is primarily spread by egg transmission. Gross lesions of the disease are manifested in lymphoid tumors in the internal organs. The liver tends to enlarge when it is affected. The use of genetically resistant strains of chickens and elimination of virus-shedding hens from the breeder flock are the best means of prevention.

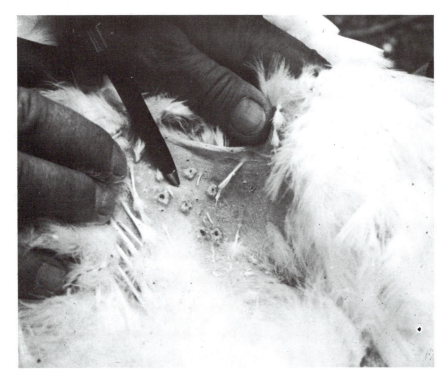

FIG. 9.5.   Skin lesions from Marek's disease. (Source: USDA.)

FUNGAL AND MOLD DISEASES

Fungi or yeast-like molds can, under certain conditions, cause major losses in poultry. Aspergillosis, commonly called brooder pneumonia, is probably one of the two most widely spread infectious mycotic diseases. This disease is caused by *Aspergillus fumigatus* spores that may occur in contaminated incubators, feed, or litter. Transmission is by inhalation of the spores from a contaminated source such as the litter. Birds show gasping and accelerated breathing, and upon posting yellow exudates are found in the lungs, air sacs, or trachea. Prevention is the best treatment. Mycotic agents are generally too expensive to treat commercial poultry.

Candidiasis, or thrush as it is sometimes called, is a mycotic disease of the digestive tract. It is caused by *Candida albicans* and results in thickened areas in the crop, proventriculus, gizzard, and vent areas. Unsanitary and overcrowded conditions can lead to the disease since it is present in the natural gut flora. Prevention is best accomplished by a good sanitation program and by avoiding overtreatment of birds with

FIG. 9.6. Typical liver tumors found in a laying hen infected with avian leukosis. (Source: USDA.)

antibiotics or other products that can alter gut flora. Treatment can be by mycostatin in the feed or copper sulfate in the water.

Favus is an uncommon myotic disease primarily of chickens and turkeys. The disease manifests itself in grayish-white crusted lesions on unfeathered skin. The disease is caused by the fungus *Trichophyton gallinae*. Mildly infected birds should be segregated to recover; badly infected birds should be destroyed.

MYCOTOXICOSIS

The toxic metabolites produced by molds growing on foodstuffs are chemical compounds known as mycotoxins. Over 120 mycotoxins have been identified that causes toxicosis in animals. Some 12 different aflatoxins have been isolated but all seem chemically similar. Aflatoxin B_1 is the most toxic and can impair immunity and cause death at low levels. Aflatoxins are produced by *Aspergillus* and frequently *Aspergillus flavus*.

One of the series of trichothecene toxins is T-2 which is produced by *Fusarium tricinctum*. It causes dermal necrosis in the mouth.

Ochratoxin A is one of several ochratoxins that may cause kidney, liver, or bone marrow damage. Ochratoxin is produced by *Aspergillus ochraceus* and *Penicillium viridicatum*.

Rubratoxin (A or B) can cause liver damage in poultry and reduced weight gains. It is produced by *Penicillium rubrum* and *Penicillium purpurogenum*. *Penicillium citrinum* produces a mycotoxin called citrinin that can cause kidney damage in young birds and acute diarrhea in layers. Ergot is a mycotoxin produced by *Claviceps purpurea*. Ergot grows on the heads of small grains such as rye. Ingestion causes degenerative changes in the heart, liver, and kidneys. Precautions should be taken in purchasing and storing grains to help prevent mycotoxicosis.

EXTERNAL PARASITES

Lice, mites, chiggers, beetles, bedbugs, and ticks (Fig. 9.7) can at times become problems for poultry. Poultry should be kept free of external parasites since they interfere with feed conversion as well as egg production, and parasite-weakened birds are more susceptible to disease. Some parasites can be carriers of disease themselves. External parasites generally lower the value of poultry by marring the skin and increasing condemnations.

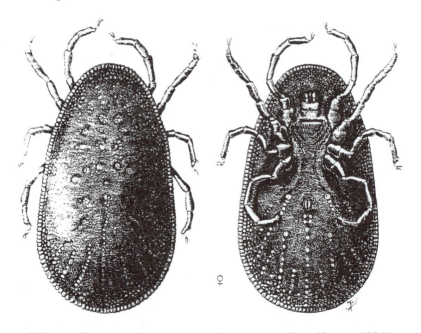

FIG. 9.7. Poultry ticks can occasionally become a problem. (Source: USDA.)

Several species of lice are found on poultry. They lay their eggs on birds, hatch, and complete their life cycle while still on the birds. Lice are generally host adapted and try to avoid light. The most common variety is the skin louse which infests the tail and vent area. To help prevent infestation, wild birds such as sparrows and starlings should be controlled and buildings should be bird proofed.

Mites are becoming more of a problem in the poultry industry as production units become larger. The chicken mite, *Dermanyssus gallinae,* or roost mite is a blood-sucking parasite that feeds on birds at night, then hides in cracks and crevices during the day. Because these mites can live approximately 6 months without feeding they can survive between flocks in an empty house. Transmission can be from bird to bird as when mating or from farm to farm carried on egg flats or cases. English sparrows are a primary vector.

The northern fowl mite, *Ornithonyssus sylarium,* commonly called the feather mite, is similar in appearance to the chicken mite. The northern fowl mite sucks blood and tissue juices from the host. A sign of infestation is the black tufting around the feathers of the vent and tail caused by deposits of egg masses. Mites cause stress in birds, which causes decreased egg production and lowering of fertility.

Other mites that can cause problems from time to time in poultry are the scaly leg mite, *Knemidoctopes mutans,* and the airsac mite, *Cytoditis nudus.* Chiggers, *Trombicula alfreduggesi,* commonly called red bugs, are common in the southeast and can be a problem primarily in range turkeys. The chiggers feed on tissue that is liquified by a substance injected at the site of the bite. In addition to irritating and stressing the birds, carcasses are downgraded at processing because of trimming to remove the sites of irritation. Prevention by carefully selecting range sites and treating range grounds when necessary is the best treatment.

The darkling beetle, *Alphabatobious diaperinus,* or lesser meal worm is common in the litter in most broiler operations. Darkling beetles attack live poultry on occasion but usually feed on dead carcasses. They are potential disease carriers and are also destructive to poultry house insulation. They serve as intermediate hosts for tapeworms and are suspected to be carriers of botulism, salmonellosis, and Marek's disease.

Bedbugs, *Cimex lectularius,* are blood-sucking parasites that feed at night. Bedbugs can survive up to a year without feeding. Anemic and unthrifty birds with signs of skin irritation are signs of infestation. Bedbugs are becoming more of a problem in the south and can often be found in broiler breeder flocks.

Fleas of all species can cause anemia and lowered egg production.

Proper sanitation and control of wild birds and rodents are good preventive measures. In the United States, the common flea varieties are the sticktight flea, *Echidnophaga gallinaceae,* the Western hen flea, *Ceratophyllus niger,* and the European chick flea, *Ceratophyllus gallinae.*

The fowl tick *(Argas persicus)* is not a true tick. The fowl tick is a flat oval reddish brown bug. These blood sucking arthropods can be a problem in the Southern and Southwestern segments of the U.S. Fowl ticks breed during warm weather and lay their eggs in cracks and crevices in the poultry house. Since the ticks feed only at night they are difficult to detect. Fowl ticks go through a larval, nymph and adult stage in their life cycle. Since all life stages can remain alive for long periods of time erradication is difficult once an infestation occurs. Fowl ticks can be transmitted from building to building on equipment. The ticks then move from the infested house onto the birds and then from bird to bird. Fowl ticks are not parasites of people. The mortality is usually not severe, highest losses can be expected when newly hatched chicks or poults are moved into infested buildings. Treatment of infested buildings requires the use of an approved wettable powder thoroughly sprayed on walls, ceilings, floors and cracks and crevices.

External parasite populations can be kept to a minimum through continuous good management practices, proper sanitation, and the careful use of approved insecticides.

INTERNAL PARASITES

The occurrence of internal parasites in poultry is almost universal. Although internal parasites seldom prove fatal they stress the host causing poorer performance, which helps other infections to gain a foothold.

Internal parasites of poultry include large roundworms, *Ascaridia galli,* inhabit the small intestine and feed on intestinal contents unattached to the walls of the intestine. The worms are more common in floor-raised birds such as pullets. Birds become infected by ingesting feces of infected birds. Depressed behavior, unthriftiness, and diarrhea are common signs of infestation. Numerous effective wormers are available including piperazine and hygromix.

Cecal worms *(Heterakios gallinarum)* inhabit the tips of the ceca. They are not economically important but are intermediate hosts to the blackhead organism, *Histomonas meleagridis,* in turkeys. Poultry can become infected by eating earthworms or infested feces. Several wormers can be used for problem flocks.

Tapeworms can be any of a variety of species of cestodes. They are flat white segmented parasites that are host specific and generally a problem of range birds. The tapeworm head attaches to the wall of the small intestine and feeds on the contents. The life cycle is indirect and dependent on an intermediate host such as earthworms, slugs, and snails. Proper disposal of manure and eliminating wet areas in which intermediate hosts accumulate are the best preventive measures. Treatment can be with di-*n*-butyl tin dilaurate followed by removal of old litter.

Capillaria worms, commonly called hair worms, can be any of four species, *Capillaria contorta, C. columbae, C. annulata,* or *C. obsignata.* Capillaria worms are blood suckers that are passed by ingesting feces from infected birds. *Capillaria annulata* and *C. contorta* are found in the crop and esophagus. *Capillaria columbae* and *C. obsignata* are found in the small intestine. Several wormers are effective. Vitamin A supplementation aids recovery of infested birds after worming.

Gapeworms, *Syngamus trachea,* are bloodsucking nematodes that frequently infect turkeys and pheasants under range-type conditions. Gapeworm ova are ingested and hatch. Later male gapeworms attach to the lining of the trachea; female worms then attach to the males forming a "Y" appearance. Suffocation of the birds may occur in heavy infestations. Signs include gasping, choking, and "gaping" to breathe. Thiabendazole used in the feed appears to be a satisfactory treatment. Good management and preventive measures can greatly lessen the incidence of infestation of internal parasites.

BEHAVIORAL DISEASES

Hysteria, a sudden and unprovoked migration to one end of the house resulting in piling up and possible suffocation, occurs in poultry flocks with losses and damage to the flock. Malnutrition, vitamin deficiencies, hereditary, and trauma pain have been suggested as possible causes. Light breeds such as leghorns are the most susceptible.

Cannibalism is a condition in which birds actually pick or eat each other to death. Prevention by debeaking and good management are the best ways to control this condition. Generally, overcrowding and overheating are associated with this condition. Excessive light intensity and certain nutrient deficiencies can also aggravate the problem.

REPRODUCTIVE DISEASES

Eversion or prolapse of the oviduct occurs in pullets and young hens. It is caused by straining during or following the laying process. Obesity in meat-type birds can contribute to the problem.

An impacted oviduct is a condition in which an egg becomes lodged in the oviduct. Internal layers are birds whose egg yolks fall into the abdominal cavity instead of the oviduct where they are reabsorbed.

Disease prevention and control programs consist of isolation, proper sanitation, good management, and a vaccination program suited to the area of production.

REFERENCES

Biester, H. E., and Schwartze, L. H. 1965. Diseases of Poultry, 5th Edition. Iowa State Univ. Press, Ames, IA.

Harris, J. R. 1965. Vaccination Can Prevent These Diseases. Poultry Science and Technology Guide No. 2, NCSU, Raleigh, NC.

Harris, J. R. 1967. Coccidiosis Control in Layers. Poultry Science and Technology Guide No. 3, NCSU, Raleigh, NC.

Harris, J. R. 1971. Sanitation and Isolation of Poultry Houses. Poultry Science and Technology Guide No. 4, NCSU, Raleigh, NC.

Harris, J. R. 1984. *Mycoplasma gallisepticum*. Poultry Science and Technology Guide No. 1, NCSU, Raleigh, NC.

Hofstad, M. S. *et al.* 1972. Diseases of Poultry, 7th Edition. Iowa State Univ. Press, Ames, IA.

Marsh, G. A. 1965. Poultry Diseases. Poultry Production in Ohio. Ohio State Univ., Columbus, OH.

North, M. O. 1984. Commercial Chicken Production Manual, 3rd Edition. AVI Publishing Company, Westport, CT.

Patrick, H., and Schaible, P. J. 1980. Poultry Feeds and Nutrition, 2nd Edition. AVI Publishing Company, Westport, CT.

Schwartz, D. M. 1977. Poultry Health Handbook, 2nd Edition. The Pennsylvania State University, State College, PA.

10

Poultry and Egg Marketing

Marketing of farm products, whether poultry, eggs, or any other commodity, involves assembling, transporting, assumption of risk, processing, storage, inspection and grading, packaging, and merchandising or retailing. Figure 10.1 illustrates the different costs in producing and processing a chicken broiler.

ASSEMBLING

Integration in the poultry industry permits the collection of large amounts of a uniform product available when needed. Birds are produced in 20,000–100,000 units within an area usually concentrated around a processing plant and feed mill. In the case of broilers, hatching and production can be scheduled to meet the needs for a particular week, 10–12 weeks ahead. In the case of broilers, turkeys, and fowl for meat, processing plants can coordinate production needs by hauling in flocks from several farms to supply the birds needed for the plant for a day.

Truckload lots of eggs are produced every day on large farms. In egg production operations this means that routine scheduled runs can be

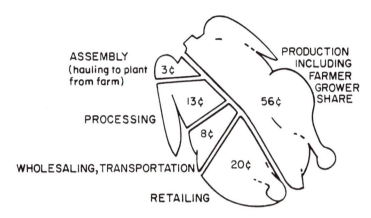

FIG. 10.1. What the consumer dollar pays for. (Source: Anonymous 1977.)

made several times a week to collect eggs from a given farm. Since the producer does not have to save eggs for a week or two to build up a truckload, the eggs, which are a semiperishable product, can be moved through market channels several days earlier and be of higher quality.

TRANSPORTING

Because production units are located close to processing plants, the costs of transporting live birds and eggs are reduced. In the case of eggs, short hauls help prevent quality deterioration resulting from jarring. In the case of live chickens and turkeys, there is less injury, mortality, and loss in yield during short runs, particularly during adverse weather conditions. Most company trucks and cars now have two-way radios, permitting constant coordination between the plant and the farm. For example, if a truck breaks down or becomes stuck in snow or mud, the processing plant has several hours to send in an alternate source of birds to keep the plant running near capacity.

ASSUMPTION OF RISK

Risk of loss is involved whenever live poultry and eggs are in transport, being processed, stored, or delivered to stores. The cost of a single trailer truckload of eggs or a house of broilers or turkeys can be as high as a fraction of a million dollars. Since live birds are subject to injury and death, processed birds are subject to spoilage, eggs are subjected to breakage and decline in quality, and all commodities are subject to declines in prices, it is not possible for producers to finance the large amount of risk involved. As a result, most integrated poultry operations are now owned by large corporations, which can more easily stand the loss of a thousand or a million birds or eggs without going bankrupt. Typical examples of such losses are condemnation of large lots of birds or processed carcasses accidentally contaminated with pesticides or harmful chemicals, extreme weather conditions such as heat waves or tornados, accidents in which trailer truckloads of eggs or broilers are completely destroyed, breakdown in refrigeration while a product is being stored, and market price fluctuations.

MARKET NEWS

The U.S. Department of Agriculture operates a market news service in cooperation with state agencies for broiler/fryers, fowl, roasters, turkeys, ducks, squabs, other miscellaneous poultry items, and rabbits. Egg reports are made on frozen and dried eggs. The reports include data on movement, supply and demand, cold storage stocks, trading activity, price activities, and quality ranges. Information on poultry and eggs is gathered by professionally trained reporters from farms, country assembly plants, shipping points, and major consuming centers.

The prices and statistics collected are disseminated nationwide by the USDA leased wire service. Other wire services, newspapers, radios, and television media as well as mailed reports are then used to disseminate the information to the primary users.

Seventeen different state offices in poultry and egg-producing areas or in terminal markets now provide market news services on poultry and eggs (e.g., *Egg Market News Report*).

FUTURES TRADING

With modern production methods, the risk of losses of birds from mortality and condemnation from diseased conditions can be fairly accurately estimated when millions of birds are involved. Because of integration, producers are guaranteed so much a pound to produce birds. However, the integrating firm still runs the risk of fluctuating prices of grain, which amounts to a large percentage of the cost of production. Fluctuations in the price they receive when broilers or eggs are ready for market also represent a large risk. To protect themselves from rapid increases in grain costs once they start to grow chickens and from a drop in price below the cost of production when birds are marketed, the companies practice "hedging" in the futures market.

Futures trading in both fresh, white shell eggs and broilers is conducted by the Chicago Mercantile Exchange. They set up the rules for trading. For example, broiler futures are handled in units of 30,000 lb and egg futures in units of 750 cases of eggs. The eggs or broilers may not even exist at the time the contract is drawn up.

Prices are locked in by hedging. The following illustrates the use of hedging in a broiler operation. Before starting broilers, the integrator, based on past experience, can calculate fairly closely the costs of production; he therefore knows the selling price necessary to make a profit. Assuming that the processed birds will be ready for market in December, December broiler futures are 48¢ a pound and he knows that at that price, he can produce dressed broilers, make a reasonable profit, and pay the costs incurred in selling a futures contract. To protect himself from a decline in broiler prices, he buys a futures contract when he sets eggs or starts his chicks, for which in effect he agrees to deliver a stated quantity and quality of dressed broilers for 48¢ a pound in December. If when he markets his broilers in December the broiler price has dropped, he may receive only say 42¢ a pound for the broilers he produced, which amounts to a 6¢ a pound loss. However, December futures would also have dropped to 42¢ a pound. To offset his cash losses from the sale of his broilers, he purchases a contract to buy broilers in December at 42¢ a pound and delivers it to the purchaser who had contracted to purchase broilers at 48¢ a pound. The 6¢ a pound he gains in the futures trading offsets his losses from the sale of his broilers.

On the other hand, if the December price had risen to 54¢ cents and he made 6¢ a pound on his broilers, he is still obligated to deliver broilers or a contract for 48¢ a pound. To fulfill his contractural obligation to deliver December broilers at 48¢ a pound, he must purchase a futures contract at 54¢ a pound to deliver broilers in December. The extra 6¢ a pound he must pay for his contract is offset by the 6¢ a pound extra he received for his broilers.

To guarantee the prices for feed, he purchases contracts before starting his broilers to feed at a certain price on a certain date or hedges corn and soybean futures contracts to help lock in feed costs.

PROCESSING POULTRY

Table 10.1 lists the functions performed in a poultry-processing operation. Chickens and turkeys are transported in removable plastic crates, palletized cages, or metal cages that are a part of the trailer truck. At the plant, birds are removed from the crates and hung on shackles attached to a conveyor line. The line moves to a bleeding tunnel in which the birds are electrically stunned and their throats are

TABLE 10.1. Poultry-Processing Functions

Assembling	Slaughtering	Eviscerating	Additional processing uncooked
Catching	Shackling	Removing preen gland	Whole carcass
Loading	Killing	Opening and inspection of remaining viscera	Whole carcass cut up
Transporting	Bleeding	Separation of heart, gizzard, and liver	Individual parts
Unloading	Scalding	Washing	Half carcasses
	Defeathering	Grading	Hand or mechanically deboned meat
	Washing	Weighing	
	Removing head and feet	Chilling	

Additional processing cooked	Packaging	Storage	Other
Cooked, deboned meat	Plastic bags with and without vacuum	Ice packed	Plant cleanup and sanitation
Fried battered parts	Tray with plastic overwrap	Chilled	Waste disposal
Barbecued	Cardboard cartons including laminates and overwraps	Dry pack, refrigerated	Waste treatment and by-product recovery
Canned whole or deboned	Metal, cardboard, or wood ice-packed bulk containers	Frozen	Insect and rodent control
Canned soups, stews, etc.	Cans	Canned	Plant and product safety
Comminuted products	Laminated pouches	Freeze or heat, dehydrated	Plant security
Rolls		Irradiated (future)	Fiscal and personnel management
Ham			

cut by hand or with a circular knife for dispatching and bleeding. The carcasses, while still shackled on the conveyor line, are carried and immersed in water of 123°–138°F (50.6°–58.9°C) for 30–120 sec to facilitate feather release.

For defeathering, a series of two to four rubber-fingered pickers are used. Each picker is designed to remove feathers from a specific part of the carcass. Any feathers remaining must be removed by hand or singed with a flame to remove hairs or "filoplumes." Finally, before transfer to the eviscerating line, which is in a separate room, the carcasses are washed and the head and feet are removed.

Blood, feathers, heads, and feet are moved to a recovery station for use as a by-product, generally a feed animal protein supplement. Some water is recycled and used in the plant in areas in which it does not come in contact with the carcasses.

The carcasses are then transferred to an eviscerating line. There the preen gland is removed, the carcasses are cut open at the vent, and a machine automatically removes the viscera, which is left hanging outside the carcass so that both the carcass and viscera can be inspected for wholesomeness by the USDA inspector. Next, the edible viscera or giblets (heart, liver, and gizzard) are removed, cleaned, chilled, and wrapped after separation from the carcass. The inedible viscera drop down into a trough or gutter, from which they are moved to the recovery area. Next, the carcasses, while still on the eviscerating line, have the crop, lungs, and other extraneous material removed by vacuum. Finally, the neck is removed and the carcass is washed, inspected, and automatically released from the shackles into a chiller.

Carcasses are cooled in circulating water chillers from about 100°F (38.1°C) to 36°F (2.2°C) in stages in about 30–45 min. Then they are removed from the chillers on conveyors and hung on a drip line to remove excess moisture. While moving on the drip line, the carcasses are automatically released by sizers, which drop them into bins or tanks according to weight classes. Finally, wrapped giblets are inserted into the carcass for packaging or the carcasses are held without giblets for additional processing. The carcasses are graded sometime after they come from the chillers. Figure 10.2 shows a flow diagram of a typical processing operation.

One of the more important recent developments in the poultry industry has been the gradual adoption, improvement, and production of a process that mechanically debones poultry meat. To produce such meat, whole carcasses, backs, and necks are ground and then processed through a deboning machine that removes bone fragments by sieving. Since such meat can be produced profitably for less than similar red meats, demand has been growing rapidly for use in mechanically deboned meat products. Chicken and turkey hot dogs are examples of products made from this meat.

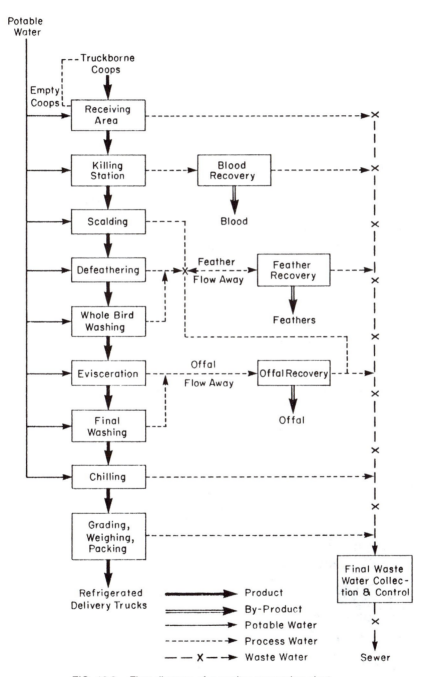

FIG. 10.2.   Flow diagram of a poultry-processing plant.

FURTHER PROCESSING

In recent years, the trend has been toward further processing of poultry carcasses (Fig. 10.3). Further processing started by simply cutting up broilers and packaging them individually in trays. Further processing is a marketing technique that provides convenience to the consumer and allows the processor to add value to the product by providing an additional service. As processors became successful in providing more services to the consumer, the product mix was expanded to include parts and eventually deboned products. Large integrators such as Holly Farms and Perdue Foods were some of the first companies to advertise their products regionally. Consumers recognized the quality of these products and were willing to pay more for brand name products in spite of the generic connotation of chicken. Further processing and packaging of poultry products by the processor appeal to retailers as well as consumers. Products now arrive prepackaged, weighed, dated, and even prepriced if the retailer desires these services. This allows the retailer to use a minimal amount of labor in handling the product, primarily in stocking the product in the meat counter.

Poultry processors have been successful in moving large volumes of prepackaged value added poultry products by offering convenience and service to both retailers and consumers. The competitive price of poultry compared to red meats has led to the acceptance of prepackaged

FIG. 10.3.   A modern poultry-processing operation.

higher value poultry items. As the demand for poultry parts increased, it created a surplus of the less desirable parts such as backs and necks. Gradually the adoption, improvement, and production of mechanically deboned raw poultry meat has taken place. In processing this meat, backs, necks, or whole carcasses are ground and the bone fragments and cartilage removed by sieving. Meat of this type can be produced more cheaply than similar red meats. Chicken and turkey hot dogs are examples of products made from mechanically deboned poultry.

The trend in further processing is for more convenience and time savings built into the product with the processor able to add value to the basic product. This trend will allow for an expanded offering of poultry products such as deboned products, partially and fully cooked items, as well as expanded institutional offerings.

EGGS

Processing of shell eggs is done mainly to maintain quality. Since eggs are small, fragile, and have a low cost per unit, they must be handled rapidly and with extreme care. Shell egg-processing machinery must be protected from egg whites and yolks from broken eggs and be easily cleaned to keep contamination at a minimum.

Generally, shell eggs are cooled and held at 60°F (15.4°C) or lower and 70% relative humidity, which are the optimum storage conditions. Quality can be maintained a little longer by supplementing the storage conditions with oil treatment of the eggshells. This is done by spraying the eggs with a mist of mineral oil as soon as they are gathered at the farm or after processing to prevent loss of carbon dioxide. Carbon dioxide loss sets in motion a complex series of chemical and physical events that lead to thin, watery whites and flat, enlarged yolks.

For processing eggs stored in 30 dozen cartons, the eggs are removed from the cooler, picked up 30 at a time by a vacuum lifter, and transferred to a rubber conveyor line on which they move through an egg washer. To prevent bacterial contamination and possible spoilage during washing, the eggs are washed in water of at least 90°F (32°C) and then rinsed in water hotter than the wash water. The wash water contains a detergent for cleaning and a sanitizer to control microorganisms. After the eggs have been washed, the shells are usually sprayed with a thin coating of oil, flash candled, and sized. They are automatically packaged in one dozen-sized cartons or bulk packed.

Some eggs are broken out and sold as liquid eggs. Figure 10.4 shows the flow of products through an egg breaking plant. Other eggs are dried.

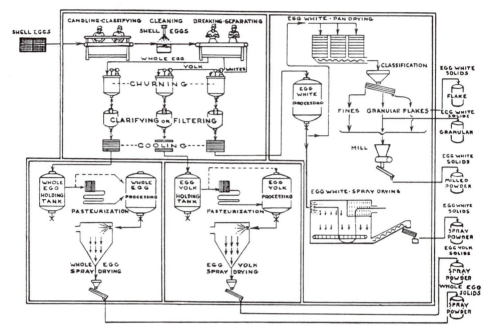

Courtesy of The Bakers Digest

FIG. 10.4. Schematic outline showing the various processing steps involved in the production of whole egg, egg yolk, and egg white solids. Source: Jewell et al. (1975).

STORAGE

Supplies of poultry and eggs for processing and marketing were highly seasonal as recently as 30 years ago. Large amounts came on the market in one season with resultant low prices and at other times of the year there was a scarcity with resultant high prices. To stabilize prices and provide a uniform supply, products were stored frozen in the case of poultry meat and either dried or in frozen liquid form or at temperatures just above freezing in the case of shell eggs. The quality of such products when removed from storage was just acceptable. With major increases in the knowledge and rapid adoption by industry of the latest scientific methods and practices of poultry production, it has been possible to provide an almost uniform supply of eggs and broiler meat every week of the year. Such a uniform supply helps eliminate the risk, cost, and decline in quality from storage.

Broilers are produced and marketed "chilled" 52 weeks a year. The turkey industry has moved to production of turkeys year round.

At present most whole turkeys are held and marketed frozen. Few shell eggs are held more than a few weeks. Chilled or frozen liquid and dried eggs, because of their uniformity and convenience, are still used in large food manufacturing operations. They help to stabilize prices because large amounts are broken when prices are low. However, because of the demand for such products and the cost of operating a breaking plant, these plants now operate all year.

INSPECTION AND GRADING

Inspection of Poultry for Wholesomeness

Because of mandatory government inspections of poultry meat for wholesomeness and voluntary and sometimes mandatory grading for quality of eggs and poultry meat, consumers have available safe, uniform, high quality poultry meat and eggs every week of the year. Although there are many state laws and quality standards usually based on federal standards, the general trend has been to centralize inspection and grading of poultry carcasses and grading of eggs at the federal level.

The federal Poultry Products Inspection Act of 1957 provides compulsory federal inspection of all poultry and poultry products entering interstate or foreign commerce. Squabs, gamebirds, and rabbits are exempt from the act but the USDA offers a voluntary inspection service for these animals when requested. Under the provisions of the act, inspectors are hired and paid by the USDA. They have the authority to remove and condemn carcasses from the processing line if they are not fit for human consumption or to stop a plant from operating when conditions are unsanitary or threaten the wholesomeness of the birds being processed. In addition, inspectors make sure products are not adulterated and that they are truthfully labeled with labels that have received prior approval from the USDA before use. Other functions include laboratory analysis of products suspected of either accidental or deliberate adulteration.

Grading Poultry for Quality

Birds can also be graded for quality by government licensed graders. Inspection for wholesomeness is mandatory; grading is voluntary. However, when a processor agrees to use the USDA grading service in his plant, he is obligated to follow all rules and regulations and pay the

USDA for the cost of the grading service. In return, the plant can display the USDA grade designation on their products and wrappers.

The USDA grades and standards applied by qualified personnel provide as accurate and objective a method for identifying quality as can be devised for practical application. Within the limits of subjective evaluation, they are the same in all areas of the country and for all conditions and types of poultry marketed. The grades and standards are a matter of public reference and are acceptable evidence in legal cases. They also serve as a basis for establishing state, trade, and research specifications. Figure 10.5 shows examples of labels bearing inspection and grade marks.

Under the grading system the different types of poultry are divided into kinds, which refer to different species of poultry such as chickens, turkeys, ducks, geese, guineas, and pigeons, and classes, which group together birds that have essentially the same physical characteristics. For example, an old rooster (cock) and a discarded laying hen (stewing chicken) differ greatly in eating qualities and amount of meat on the carcass when cooked and eaten. Table 10.2 lists the several classes of poultry.

Before applying a grade to a poultry carcass, the carcasses are sorted on the basis of condition. Any carcass exhibiting protruding pinfeathers, bruises, improper trimming, inedible viscera, attached or unremoved feathers or blood, feces, or other extraneous material is classi-

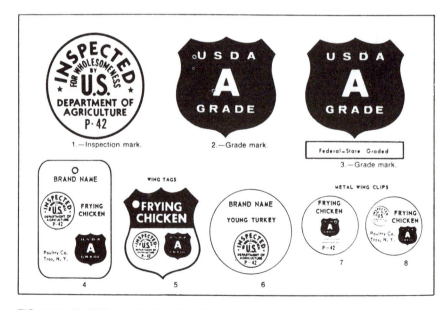

FIG. 10.5. (1–3) Forms of official identification for ready-to-cook poultry; (4–8) examples of labels bearing inspection and grade marks. (Source: USDA.)

TABLE 10.2. Market Classes of Poultry

Species	Class	Sex	Age
Chicken	Cornish game hen	Either	5–7 weeks
	Broiler or fryer	Either	9–12 weeks
	Roaster	Either	3–5 months
	Capon	Unsexed male	Under 18 months
	Stag	Male	Under 10 months
	Hen or stewing	Female	Over 10 months
	Cock	Male	Over 10 months
Turkey	Fryer-roaster	Either	Under 16 weeks
	Young hen	Female	5–7 months
	Young tom	Male	5–7 months
	Yearling hen	Female	Under 15 months
	Yearling tom	Male	Under 15 months
	Mature or old	Either	Over 15 months
Duck	Broiler duckling	Either	Under 8 weeks
	Roaster duckling	Either	Under 16 weeks
	Mature or old	Either	Over 6 months
Goose	Young	Either	
	Mature or old	Either	
Guinea	Young	Either	
	Mature or old	Either	
Pigeon	Squab	Either	
	Pigeon	Either	

[a] Source: Anonymous (1977).

fied as "no grade" and, unless spoiled, is sent back for additional washing or processing. The carcasses are then graded into three general quality designations: U.S. Grade A, B, or C.

The criteria for establishing a grade consist of the conformation or general overall shape of the carcass, fleshing or the amount of meat on the bones, fat covering under the skin, and the presence or absence of specific defects. The presence of defects such as pinfeathers, exposed flesh, cuts, tears, and broken bones or skin discolorations and flesh blemishes and bruises all influence the grade assigned to a particular carcass (see Table 10.3). Figures 10.6 and 10.7 show photographs of carcasses graded A and C.

Grading Eggs for Quality

According to the USDA, grading generally involves sorting of products according to quality, size, weight, and other factors that determine the relative value of the product. United States grades for individual shell eggs are AA, A, and B (Fig. 10.8). The United States Standards of Quality for individual shell eggs are determined on the basis of interior quality factors such as the condition of the white and yolk and the size of the air cell and exterior quality factors such as the cleanliness and soundness of the shell. Eggs are also classified accord-

TABLE 10.3. Summary of Specifications of Quality for Individual Carcasses of Ready-to-Cook Poultry and Parts Therefrom[a]

Factor	A quality	B quality	C quality
Conformation			
Breastbone	Normal Slight curve or dent	Moderate deformities Moderately dented, curved, or crooked	Abnormal Seriously curved or crooked
Back	Normal (except slight curve)	Moderately crooked	Seriously crooked
Legs and wings	Normal	Moderately misshapen	Misshapen
Fleshing	Well fleshed, moderately long, deep and rounded breast	Moderately fleshed, considering kind of class, and part	Poorly fleshed
Fat covering	Well covered—especially between heavy feather tracts on breast and considering kind, class, and part	Sufficient fat on breast and legs to prevent distinct appearance of flesh through the skin	Lacking in fat covering over all parts of carcass
Pinfeathers			
Nonprotruding pins and hair	Free	Few scattered	Scattering
Protruding pins	Free	Free	Free

Exposed flesh[b]

Carcass weight		A quality			B quality			C quality
Minimum	Maximum	Breast and legs	Elsewhere	Part	Breast and legs[c]	Elsewhere	Part	
None	1½ lb	None	¾ in.	Slight trim on edge	¾ in.	1½ in.	Moderate amount of the flesh normally covered	No limit
Over 1½ lb	6 lb	None	1½ in.		1½ in.	3 in.		
Over 6 lb	16 lb	None	2 in.		2 in.	4 in.		
Over 16 lb	None	None	3 in.		3 in.	5 in.		

	A Quality	B Quality	C Quality
Discolorations[d] (breast and legs / elsewhere)			No limit[e]
None – 1½ lb	¼ in. / ½ in.	½ in. / 1 in.	½ in. / 2 in.
Over 1½ lb – 6 lb	¼ in. / 1 in.	1 in. / 2 in.	1 in. / 3 in.
Over 6 lb – 16 lb	½ in. / 1½ in.	1½ in. / 2½ in.	1½ in. / 4 in.
Over 16 lb – None	½ in. / 2 in.	2 in. / 3 in.	1½ in. / 5 in.
Disjointed bones	One	Two disjointed and no broken or one disjointed and one non-protruding broken	No limit
Broken bones	None		No limit
Missing parts	Wing tips and tail[f]	Wing tips, second wing joint, and tail; Back area not wider than base of tail and extending half way between base of tail and hip joints	Wing tips, wings, and tail; Back area not wider than base of tail extending to area between hip joints
Freezing defects (when consumer packaged)	Slight darkening over the back and drumsticks; few small ⅛ in. pockmarks for poultry weighing 6 lb or less and ¼ in. pockmarks for poultry weighing more than 6 lb, occasional small areas showing layer of clear or pinkish ice	Moderate dried areas not in excess of ½ in. in diameter; may lack brightness; moderate areas showing layer of clear, pinkish, or reddish colored ice	Numerous pockmarks and large dried areas

Source: USDA (1983B).

[a] Minimum requirements and maximum defects permitted.

[b] Total aggregate area of flesh exposed by all cuts and tears and missing skin, not exceeding the area of a circle of the diameters shown.

[c] A carcass meeting the requirements of A quality for fleshing may be trimmed to remove skin and flesh defects, provided that no more than one-third of the flesh is exposed on any part and the meat yield is not appreciably affected.

[d] Flesh bruises and discolorations such as blue back are not permitted on breast and legs of A quality birds. Not more than one-half of total aggregate area of discolorations may be due to flesh bruises or blue back (when permitted), and skin bruises in any combination.

[e] No limit on size and number of areas of discoloration and flesh bruises if such areas do not render any part of the carcass unfit for food.

[f] In ducks and geese, the parts of the wing beyond the second joint may be removed, if removed at the joint and both wings are so treated.

FIG. 10.6. Examples of USDA Grade A broiler carcasses.

ing to weight and sorted into white, brown, or mixed exterior shell colors, although shell color is not considered to be a grading factor.

Table 10.4 shows the United States weight classes for consumer grades of shell eggs. The grade of an egg is determined by the lowest individual quality factor of a particular egg. In determining quality, factors are generally divided into exterior quality, which one can readily see by visual examination, and interior quality, which can be determined only by inspection in front of a light or by breaking the egg out. Inspection of the egg in front of a light is called candling.

The external factors for which eggs are graded include the, shape, soundness, and cleanliness of the shell. Small or hairline cracks in shells can be determined only by candling or by a process called

TABLE 10.4. United States Weight Classes for Consumer Grades of Shell Eggs

Size or weight class	Minimum net weight per dozen (oz)	Minimum net weight per dozen (lb)	Minimum weight for individual eggs at rate per dozen (oz)
Jumbo	30	56	29
Extra large	27	$50\frac{1}{2}$	26
Large	24	45	23
Medium	21	$39\frac{1}{2}$	20
Small	18	34	17
Peewee	15	28	—

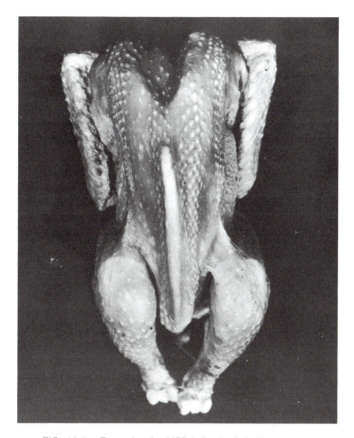

FIG. 10.7. Example of a USDA Grade C broiler carcass.

"belling," in which two eggs are gently tapped together. Cracked eggs emit a dull sound compared to eggs with sound shells. Eggs that have unusual shapes, ridges, rough areas, or thin spots are placed in grades lower than eggs with "normal" exterior shells. At the same time, eggs with stained or dirty shells are also removed before candling. Shells that have "moderate stains" are those with less than one-fourth of the shell stained.

Interior quality is determined by candling. The large end (air cell) of the egg is held before a hole with a light inside so the light illuminates the interior of the egg. To determine viscosity of the white and the shape and intensity of the yolk shadow, the egg is given a quick twirl so the contents spin. The motility of the yolk is an indication of the thickness or viscosity of the white, which becomes thinner as the quality of the egg declines. An egg with thin white has a yolk that spins easily and has a sharp or well-defined shadow or outline.

Certain defects that show upon candling such as blood clots and

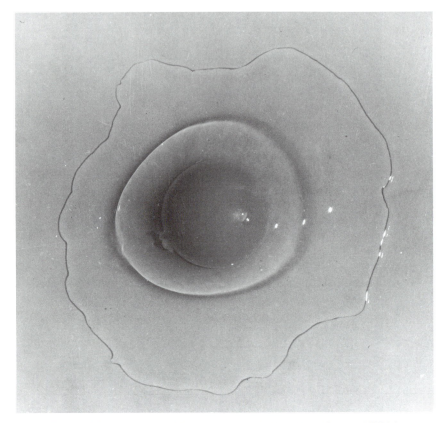

FIG. 10.8A.   U.S. Grade AA egg broken out of its shell. (Source: USDA.)

spots $\frac{1}{8}$ in. in diameter or larger, rots, molds, eggs with embryos or blood, eggs emitting bad odors, and eggs with broken shells with leaking contents result in the eggs being classified as loss eggs; these cannot receive a grade or be used for human consumption. Table 10.5 summarizes the United States Standards for Quality of individual shell eggs.

PACKAGING

The majority of broilers are sold chilled. A large portion of the chilled poultry is market packed in ice (wet packed) or dry ice in fiberboard containers. Broilers can also be packed refrigerated dry in individual bags or on pulpboard trays with a transparent plastic film wrap. Cut up carcasses are packaged in plastic foam or pulp trays

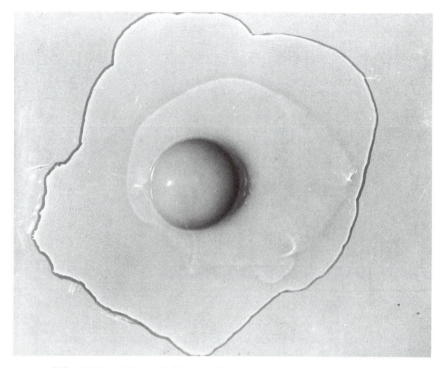

FIG. 10.8B.   U.S. Grade B egg broken out of its shell. (Source: USDA.)

overwrapped with plastic films. Cooked and uncooked frozen parts are packed in containers before freezing. Some containers may be used directly in the oven for cooking. Other forms of poultry are packaged in cans and laminated pouches. Turkeys are packed in heat-shrinkable transparent plastic bags before freezing. First, the air is exhausted from the bag and then the packaged carcass is dipped in hot water to shrink the bag. Then the packaged turkeys are rapidly frozen.

In the past eggs were usually gathered in plastic or plastic-coated wire baskets so that air could circulate freely among the eggs for rapid chilling. They are now packaged in fiberboard flats that hold 30 eggs per flat as they come off the egg-gathering belts. These flats fit in 30-dozen fiberboard cartons, 15 dozen to a side. After processing, sorting, and grading, they can again be placed on flats and packed in 30-dozen size containers or packed in 12-egg size retail cartons.

Liquid and frozen eggs are usually packaged in 30 lb (13.6 kg) tinned metal cans or plastic containers for freezing. Milk carton-type containers of 3–8 lb (1.5–3.5 kg) capacity are being used in increasing numbers.

TABLE 10.5. Summary of United States Standards for Quality of Individual Shell Eggs

Quality factor	AA quality	A quality	B quality
Shell	Clean Unbroken Practically normal	Clean Unbroken Practically normal	Clean to slightly stained[a] Unbroken Abnormal
Air cell	$\frac{1}{8}$ in. or less in depth Unlimited movement and free or bubbly	$\frac{3}{16}$ in. or less in depth Unlimited movement and free or bubbly	Over $\frac{3}{16}$ in. in depth Unlimited movement and free or bubbly
White	Clear Firm	Clear Reasonably firm	Weak and watery Small blood and meat spots present[b]
Yolk	Outline—slightly de- fined Practically free from defects	Outline—fairly well de- fined Practically free from defects	Outline—plainly visible Enlarged and flattened Clearly visible germ development but no blood Other serious defects

Dirty[c]	Check[c]
Unbroken; adhering dirt or foreign material, prominent stains, moderate stained areas in excess of B quality	Broken or cracked shell but membranes intact, not leaking[d]

Source: USDA (1983A).
[a] Moderately stained areas permitted ($\frac{1}{32}$ of surface if localized, or $\frac{1}{16}$ if scattered).
[b] If they are small (aggregating not more than $\frac{1}{8}$ in. in diameter).
[c] For eggs with dirty or broken shells, the standards of quality provide two additional qualities.
[d] Leaker has broken or cracked shell and membranes, and contents leaking or free to leak.

MERCHANDISING

Uncooked chicken and turkey carcasses are merchandised as whole carcasses or cut into individual parts, such as breasts, thighs, drumsticks, wings, and as half carcasses. Additional cooked processed products such as rolls, ham, hot dogs, bologna, loaves, and other products are made from deboned meat. Frozen, thaw, heat, and serve items such as chicken or turkey pies, or fried or barbecued chicken are also available. Other processing can be done for canned whole or deboned poultry or soups and stews. Some poultry meat is dried and used in chunk or powdered form.

Eggs are sold in shell form, dried, or dehydrated or as liquid or frozen eggs. Liquid and frozen eggs are available as plain egg whites, whole eggs, egg yolks, and salted or sugared whole eggs and egg yolks. Some egg products have other ingredients added such as corn syrup, nonfat milk, gums, whipping aids, stabilizers, vegetable oils, or other combinations of special ingredients to improve their functional properties.

FIG. 10.9.   Examples of new innovative poultry products

FIG. 10.10.   Chicken chunks or nuggets have become a popular poultry menu item.

NEW PRODUCT DEVELOPMENT

Mechanical deboning of poultry meat has made it possible to use large amounts of shredded poultry meat in products in place of red meats, at lower cost. It also means that because the less desirable parts such as backs, wings, and necks can be used for this purpose the more desirable parts such as breasts and thighs can be sold at lower costs. Figures 10.9 and 10.10 are examples of some new products recently placed on the market.

REFERENCES

Anonymous. 1977. People on the Farm: Broiler Growers. U.S. Dept. of Agriculture, Office of Governmental Affairs, Washington, DC.

Mountney, G. J. 1976. Poultry Products Technology. AVI Publishing Company, Westport, CT.

Jewell, W. J., Davis, H. R., Johndrew, O. F., Jr., Loehr, R. C., Siderewicz, W., and Zall, R. R. 1975. Egg Breaking and Processing Control and Treatment. EPA 660/2-75-019. Environmental Protection Agency, Corvallis, OR.

Mountney, G. J., and Gould, W. A. 1987. Practical Food Microbiology and Technology. Van Nostrand Reinold, New York.

Stadelman, W. J., and Cotterill, O. J. 1973. Egg Science and Technology. AVI Publishing Company, Westport, CT.

USDA 1983A. Egg Grading Manual. Agricultural Handbook No. 75, Food Safety and Quality Service, Washington, D.C.

USDA 1983B. Meat Grading Manual. Agricultural Handbook No. 31, Food Safety and Quality Service, Washington, D.C.

11

Broiler Production

The United States broiler industry evolved over a period of many years starting with the marketing of young, live surplus cockerels from laying stock in the early summer as fryers or springers. Consumers soon learned that if they purchased a white feathered bird it was probably a White Leghorn whereas if they purchased a bird with barred feathers similar to a Barred Plymouth Rock, it would be superior in meatiness and conformation to a leghorn.

Scattered small-scale production of fryer chickens started in the late 1800s and early 1900s in various parts of the United States. These were often sold in lots of less than 25 birds. It is generally recognized that Mrs. Wilmer Steele of Ocean View, Delaware started the modern broiler industry when she started 500 chickens in 1923 and sold them as fryers when they reached a weight of 2 lb. Similar interest in raising young chickens for meat purposes occurred in Georgia and Benton County, Arkansas. The early industry gained support in the area known as Delmarva (Delaware, Maryland, and Virginia) between the close of World War I and the start of World War II. Because of the demand for young, tender-meated birds, producers began to attempt to raise broilers year round. However, large-scale commercial production was limited until complete balanced rations were developed.

Before balanced rations were developed, broilers were allowed to graze on green vegetation and absorbed sunlight for vitamin D synthesis. To capitalize on the barred feathering symbol, producers began to cross the dual purpose Barred Plymouth Rock males, which were slow feathering, with the dual purpose New Hampshire females, which were rapid feathering, vigorous, fast-growing birds. The resulting heterosis or hybrid vigor, along with the fact that both sexes had barred feathering, produced an excellent carcass for that time. The opposite cross, New Hampshire males with Barred Plymouth Rock females produced males with barred feathering and females with black feathers for production of brown table eggs for the northeastern market. In Arkansas, another early broiler-producing area, the White Wyandotte was used as the stock for broiler production. The New Hampshire was also used in some areas as broiler stock.

The rationing of red meats but not poultry during World War II gave the poultry industry a rapid spurt in growth. At that time a broiler enterprise could be producing meat in 10–12 weeks and eggs in about 28 weeks. Producers could raise all the broilers they wished and consumers could purchase them without ration coupons. From that time on poultry gradually became an item in the American diet every day of the week instead of just on Sunday.

Since young chickens were being raised year round for meat purposes the terms "springers," "squab broilers," "asparagus broilers," and "hothouse chickens" really no longer applied; therefore, after World War II, as the industry evolved, these chickens began to be called broilers. The term broiler goes back to the turn of the century and literally means a young chicken that can be broiled, fried, baked, or broasted.

The broiler industry next made improvements in the meat qualities of broilers. Some small breeders kept trying to refine the dual-purpose breeds of chickens. In California, the Vantress family utilized the dark Cornish because of its superior conformation and better meat yield. The color of the Cornish was changed to red and a one-directional cross was established using a red Cornish male × New Hampshire female. Later when the first Chicken of Tomorrow Contest was held, the Vantress strain cross won and the modern broiler industry was launched. Finally, the need for white-plumaged birds for processing was realized and the red Cornish was changed to a White Cornish (Fig. 11.1A, B). Unfortunately, compared to the competition, the Vantress female line lacked reproductive performance. To compensate for this problem, the broiler industry split into male and female line breeders. The male line was dominated in early years by Charles Vantress, who by then had moved to Georgia. The female lines were fairly well dominated by the Hubbard, Cobb, and Arbor Acres breeding organiza-

FIG. 11.1A. Example of a Hubbard male broiler parent line.

tions, which used primarily a white Plymouth Rock-type hen. Unlike the table egg breeder industry, broiler breeders never established a franchise distribution system. They sold direct to the customer and supplied the birds through area distribution centers. Currently, breeders of broiler stock sell a package of male line and female line birds that the integrator can utilize to produce hatching eggs.

Once the numerous nutritional problems involved in growing chickens throughout the year were solved and the industry had standardized broiler stock, diseases became the limiting factor.

FIG. 11.1B.   Example of an Arbor Acres female line for broiler production.

During the late 1960s and early 1970s lines of pleuropneumonia-like organism-free chicks were established. Marek's disease vaccine was developed and used by the industry and a number of new and more effective coccidiostats were developed and used.

Table 11.1 shows the gains in production efficiency during the last 10 years and forecasts the gains predicted in the next 10 years based on a survey by the National Broiler Council of 10 breeders of broiler stock.

TABLE 11.1. Trends in Production Efficiency[a]

	1973	1983	1993[b]
Feed conversion (lb feed/lb meat)	2.11	2.01	1.87
Market age (days)	57	49	42
Market weight (lb)	3.88	4.12	4.34
Caloric conversion	3130	2930	2750
Hatchability (percentage usable eggs hatched vs. number of eggs set)	85	85	84
Livability (percentage market age of birds compared to 100-day-old chicks placed)	96.6	96.6	97.0
Breeder hen lay (settable eggs per broiler during productive life)	156	160	162

[a] Source: National Broiler Council.
[b] Predicted values.

UNITED STATES BROILER INDUSTRY

The United States broiler industry developed over a relatively short period of time, mainly in the southeastern United States and the Delmarva area. The industry tended to concentrate in areas in which alternative agricultural income sources were needed. Other major factors influencing the location of broiler production were access to economical sources of feed supplies, mild climate, and relatively low wage rates. Broiler production in the United States in 1983 reached a record 4.18 billion broilers with the top three broiler-producing states being Arkansas, Georgia, and Alabama, accounting for nearly 43% of the total United States production. Table 11.2 shows the top 10 broiler-producing states. Recently, consumption of broiler meat in the United States increased to the point at which it now approaches 50 lb per capita.

In 1982 broilers posted cash receipts of $6,780,100,000 thereby, ranking seventh in all United States farm commodities. It appears that United States broiler production can be expected to expand at a rate of about 3% a year as it competes favorably with the more expensive red meats.

WORLD BROILER INDUSTRY

Broiler production is increasing in all parts of the world. The rapid generation of broilers as well as the adaptation of broilers to mass production techniques have been keys to increased production throughout the world. Since the United States produces over one-third

TABLE 11.2.   Broiler Production by Leading States, 1986:
Numbers Produced (Millions)[a]

State	1986
Arkansas	786.6
Georgia	697.4
Alabama	587.5
North Carolina	450.5
Mississippi	335.7
Maryland	263.9
Texas	238.6
Delaware	196.8
California	184.8
Virginia	154.2

[a] Source: USDA (1987).

of the broilers, the major source of poultry meat worldwide, one might
expect the United States to outdistance other countries in per capita
consumption. However, as shown in Table 11.3, the consumption of
poultry meat is relatively high in several other countries, perhaps
reflecting some avoidance of red meats such as pork.

Demand for poultry products, particularly whole broilers, has been
active in world trade in recent years. The United States has not been
competitive in the export of broilers because of the strong United
States dollar and subsidized competition from Brazil and the European
community. The leading exporters of broiler meat are shown in Table
11.4.

France and Brazil combined export over two-thirds of all exported
broilers. The United States exports the majority of its whole chickens
to Mexico, Japan, Singapore, and Canada. Cut up chickens and parts
were exported in quantity to Japan, Singapore, Hong Kong, and
Jamaica accounting for over $210 million of the $428 million of United
States poultry exports in 1983.

World demand for poultry meat, namely broilers, apparently leveled
off in 1983. Increased self-sufficiency in markets such as Saudi Arabia
and Kuwait, reduced imports by other countries because of fiscal re-
straints, and the general slowdown in the world economy brought
about at least a temporary stagnation in global trade in poultry products. Recent bids on Middle Eastern tenders for whole broilers have
been in the $1000–$1140 per metric ton range, with Brazil and France
dominating the market. In 1984, exports of United States poultry,
eggs, and products are projected to show moderate growth (5%), nearly
all of it based on the anticipation that broiler parts will bring higher
prices.

TABLE 11.3. Per Capita Poultry Meat Consumption:
Leading Countries, 1983[a]

Country	Kilograms
Israel	41.4
United States	29.8
Saudi Arabia	28.1
Hong Kong	26.7
Kuwait	26.0
Canada	22.7
Spain	21.4

[a] Source: USDA.

It is projected that world poultry consumption will increase 30% over the next decade with production in the developing countries expanding by over 50%.

BROILER BREEDERS

Broiler breeders are the parents or hatching egg producers from which commercial broiler chicks are secured. Genetically speaking, it appears that most male line broiler parent stocks are pure lines. The female line broiler parents are usually a two-way cross. Commercial broiler chicks are often referred to as terminal crosses because all of the offspring are slaughtered; this is because the cross creating them makes them heterozygous, expressing so-called hybrid vigor for market production. This makes them unsuitable as breeders because, like hybrid corn, they will not breed true. Commercial broiler chicks perform satisfactorily for the grower and contractor but have a certain amount of built-in obsolescence since the parent stock must be replaced with similar parent stock from the breeder.

Quality broiler breeding stock is available from about 7 United States companies and several foreign countries in the United States.

TABLE 11.4. Broiler Meat Exports: Leading Countries, 1983[a]

Country	1000 metric tons	Percentage
France	370	38
Brazil	289	29
Netherlands	220	22
United States	196	20
World total	981	100

[a] Source: USDA.

These same companies tend to supply much of the world demand for breeding stock.

Recently in the United States large integrators have started breeding programs to supply at least part of their own breeder stocks. Integrated companies working with breeding programs include Perdue Foods, Holly Farms, Tyson's Foods, and Conagra.

Broiler breeding stock can be purchased as a package or male line or female lines of choice. Most broiler integrators use a package program from a breeding company.

Breeding stock is generally purchased as day-old chicks. Normally, an integrator purchases only 20% as many males as females. This allows enough extra males for some selection and provides adequate males to provide for a mating ratio of 1 male to 10 females.

Young breeder stock can be raised with the sexes separated or together; however, most firms rear them mixed. Obesity is a problem with breeder stock. For this reason it is an industry practice to limit or restrict the feed on breeders after approximately 6 weeks of age to prevent them from becoming obese. Overweight breeders perform at lower levels and are less heat tolerant than restricted breeders.

HOUSING AND EGG PRODUCTION

Broiler breeders are normally housed at 20–22 weeks of age and light stimulated at 22 weeks of age. This means egg production should start at about 24 weeks of age. Normally, broiler breeder flocks are kept in production from 36–44 weeks depending on the demand for hatching eggs, production level, and fertility and hatchability of eggs from that flock. A benchmark for broiler breeder hen production is 160

FIG. 11.2.   A broiler breeder house. Courtesy of Perdue, Inc.

eggs per hen and 120 broiler chicks per hen housed (Fig. 11.2). In the United States, broiler breeders are normally housed on partial slat floors. A house of this design has slats over two-thirds of the floor area in which the feeders and waterers are located. The remaining one-third of the floor area has shavings or a similar material for litter and serves as an activity area for mating, dusting, and scratching. The slats are placed 24 in. off the floor with the nests arranged so the hens can go to the nests directly from the slats. Broiler hens are normally housed at a density of 2.0 ft.^{2} per hen housed. Broiler breeder hens and roosters have their feed restricted while in production to about 80% of a full ration. Partial slat floors are utilized because cleaner hatching eggs can be produced in this fashion. Total slats tend to cause increased incidence of foot and leg problems in males.

CARE OF HATCHING EGGS

Hatching eggs should be gathered at least 3 times a day and cooled below physiological zero as soon as possible. Egg-holding rooms should be held at 55°–65°F (13°–18°C) with a relative humidity of 65%. Since broiler hatching eggs are normally not washed before setting they should be fumigated at the farm. When gathering hatching eggs they should be sorted and undesirable eggs should be separated from the hatching eggs, which are stored small end down.

THE HATCHERY

Hatching eggs normally are collected from the farm twice per week where they are identified by flock number and set. Broiler chicks are normally removed from the hatcher on the morning of the twenty-second day.

HOUSING FOR BROILERS

Broiler houses are generally designed in what has sometimes been described as a modified "A" frame design with widths of 32–50 ft wide (29–45 m). The majority of the new houses in the southeast are 40 ft wide (36 m). At this width the houses have a clear span (no posts) and still make use of some natural ventilation. The length of broiler houses varies but most new houses are 360–500 ft (324–450 m) in length although a few 800-ft (720 m) houses are in operation. Broiler houses are curtain sided and have dirt floors. All broiler houses built today are insulated in the ceilings and ends with most houses insulated in the sidewalls as well (Figs. 11.3 and 11.4).

FIG. 11.3.   Interior of a modern broiler breeder house. Courtesy of Perdue, Inc.

LITTER

Broiler house floors are covered with 3–4 in. (7.5–10.0 cm) of litter material over a dirt floor. Materials such as wood shavings, rice hulls, peanut hulls, ground corn cobs, and shredded paper are used for litter. Regardless of the material used, it should be absorbent, inexpensive, readily available, and not create problems for the birds or for use as manure.

Litter materials in broiler houses are now normally utilized to grow several groups of broilers. The trend in the industry is to completely clean out the house at least once every 2 years. When end room brooding is used, houses are cleaned in the brooding area at least once a year. Since built-up litter holds heat it is good management to clean the houses in spring and utilize the heat production of the litter in winter to help keep the broilers warm while saving brooder fuel.

BROODING

Brooding is the time period in which supplemental heat is supplied to broilers to help them maintain body temperature and grow at a

FIG. 11.4. A modern efficient narrow curtain broiler house. Courtesy of Perdue, Inc.

more optimal level. Supplemental heat for broilers is normally supplied by a hover-type brooder that has the heat unit covered with a rounded piece of sheet metal to deflect the heat downward. These brooders are usually suspended from the ceiling with light cable and pulleys so they can be raised and lowered and drawn up out of the way after brooding.

Several types of brooders are available to equip broiler houses. Flat-top or pancake-type hovers are popular. They utilize a special burner that produces radiant heat. They are about 4 ft (3.6 m) in diameter and for brooding are suspended about 2 ft above the litter. Usually 500 to 750 chicks are placed per brooder. Catalytic brooders are also available that use a catalyst to produce a chemical reaction for heat production. These brooders produce a clean, flameless heat and they are not affected as much by dust and moisture as other types. The surface temperature of the heating device is low and will not ignite dust or litter. The heat is deflected from the burner so no hover is used and fuel consumption is less than conventional gas brooders.

BROODING METHODS

In recent years higher costs for all fossil-based fuels have resulted in higher brooding costs. Since broiler houses are curtain sided much of the brooder heat produced is lost. In an attempt to lower brooding costs several innovative methods of early brooding have been developed, such as "end room brooding," "partial house brooding," and others. Regardless of the name, the goal is to brood broiler chicks more economically. For example, during the first 3 weeks of brooding the chicks are housed in only a portion of the house so less space is heated and some heat generated by the chicks themselves is used. Later, the chicks are allowed to migrate to other sections of the house. Partial house brooding can save considerably on the brooding expense but requires good management and proper ventilation to prevent ammonia buildup and wet litter.

DENSITY

Density is the amount of space allowed per bird in the house. Since housing costs are high, integrators attempt to place broilers at relatively high density but provide enough space so performance is not compromised. Generally speaking, broilers are placed at a density of approximately 0.75 ft^2 per bird (14 birds per m^2). Table 11.5 gives the placement headcount for some various sized broiler houses.

Bird density is influenced by the time of year, amount of insulation in the house, mechanical ventilation, and whether the houses are equipped with fogging equipment in locations where temperatures and humidity are high.

PLACEMENT

Normally 3 to 5 days prior to chick placement, the contract grower and service person from the integrated firm review programs and

TABLE 11.5. Placement Headcount for Broilers[a]

House dimensions (ft)	Large broiler[b]	Small broiler[c]
32 × 400 (9.6 × 120 m)	16,700	20,500
36 × 360 (10.8 × 108 m)	16,700	20,500
40 × 400 (12 × 120 m)	20,800	25,600
40 × 500 (12 × 150 m)	26,000	32,000

[a] Source: Harris (1964).
[b] Large broiler—4.0–4.5 lb live weight.
[c] Small broiler—2.5–3.2 lb live weight.

conditions for placement of chicks. At that time the service person determines whether the house or houses are ready for placement. The houses should be checked again with special care on the first day to determine if chicks are comfortable and again on the third day.

MANAGEMENT SUGGESTIONS FOR PARTIAL HOUSE BROODING[1]

1. Regulate the room temperature to 85°F (30°C) at the feed line level.
2. Place the suspended automatic feed lines down.
3. Place the water jars and feed trays in straight lines to assure good distribution.
4. Adjust the height of the brooder to approximately 36 in. (91 cm) from the litter.
5. Use timed fans to remove ammonia; if temperature cannot be maintained at 85°F (30°C) turn up the brooders.
6. Place the automatic waterers down after the chicks are placed.
7. Use adequate light, starting with 75- or 100-W bulbs.
8. Use a standard lighting program: 23 hr of light and 1 hr of darkness. Normally, chicks are kept 2.5 weeks in winter in the brooding area and only 11 to 12 days in summer before opening up more of the house.
9. Adjust brooders to maintain 80°F (65°C) in the brooding area and expanded area of the house.
10. Allow birds to move to the expanded housing area on their own (do not force them).
11. Use larger light bulbs in the expanded area to attract the broilers.
12. Utilize sufficient fans to remove ammonia and maintain good litter conditions.
13. Lower the temperature of the house to 75°F (24°C) at 4 weeks of age. Where fan ventilation is sufficient, optimum weight gain and the best feed efficiencies occur when the broilers are kept in this temperature range.
14. Temperature should remain at 75°F (24°C) until 1 week before processing, then lower it to 70°F (21.5°C).
15. Utilize sufficient fans to keep ammonia levels low and maintain good litter conditions to prevent manure burns.

[1]Offered by several integrated firms in North Carolina.

WATERING EQUIPMENT

Clean fresh water is essential for a maximum rate of gain and should always be available. Several delivery systems can be used for watering broilers. Often two systems are used together to start broilers since good water management is essential to quality and performance. The following practices are generally used.

Water Troughs

1. Water must be level end to end.
2. Troughs should be washed daily and not dumped into the litter.
3. Water depth ¾ in. (first joint of your index finger), 2 cm.
4. Maintain the height so the trough bottom is even with the back of the average broiler.

Low-Pressure Drinkers

1. Check pressure gauge to make sure water will flow properly.
2. The water depth should be ½ in. (top of the fingernail on the index finger), 1.25 cm.
3. Maintain the height so the bottom of the drinker is even with the back of the average broiler.

Water Jars or Jugs

1. Supplemental 1-gal (3.8 liter) water jars may be used the first few days; they must be level.
2. Cleaned daily and never empty.
3. Gradually start removing them by the fifth or sixth day.
4. Use 6 supplemental waterers per 1000 broilers.

FEEDING EQUIPMENT

Availability of feed is important. For starting chicks most broiler houses use pan-type automatic feeders supplemented by feeder lids made of either reusable plastic or discardable cardboard (Fig. 11.5). In either case, the lids are approximately 16 × 24 in. (40 × 60 cm) in size and are placed on the litter. Usually one feeder lid is used per 100 chicks and about 2 lb (1 kg) of feed is scattered per lid. They should be refilled as needed and removed by 10 days of age.

Trough feeders are not used very much in the United States for

FIG. 11.5. Healthy chicks started on automatic feeders with pine shavings for litter. Courtesy of Chore-Time, Inc.

broilers because of high labor requirements, since it requires 25–30 5-ft (1.5 m) troughs per 1000 broilers.

Mechanical chain feeders can be used in broiler houses. Roughly 140–160 linear ft (43–49 m) of feeder space is required per 1000 broilers.

When 30 lb capacity (14 kg) tube-type feeders are used it takes 25–30 per 1000 broilers.

Suspended round pan feeders (automatic centerless auger feeders) are popular among broiler producers. Generally, 2 lines are used in houses less than 36 ft (11 m) and 3 lines in houses over 36 ft in width. Wire grills in the pans prevent wastage of feed and electro guards are used to prevent roosting on the feeders. Generally, with this system, a broiler needs about 3 in. (7.5 cm) of feeder space per bird. Regardless of the feeding system used, feeders should be adjusted so the lip of the feeder is even with the back of the average broiler.

Bulk feed bins are used to store broiler feed. Generally, these bins hold about 8 tons (7500 kg) each and are built to receive delivery in bulk by auger filling or by use of air. A rule of thumb is that storage capacity should be an 8 to 10 day supply at maximum usage. It is important that bins be waterproof to prevent the formation of aflatoxins in molding feed.

LIGHTING PROGRAMS

Broilers are generally grown on a continuous light program. Many integrators recommend a schedule of 23 hr of light to 1 hr of darkness in a 24-hr cycle to accustom the chicks to darkness in case of a power failure. Relatively high levels of light are used initially to help the chicks find food and water. Starting at about 12 days of age, the level of lighting is reduced to 25–40 W per 1000 broilers. This level is generally used throughout the remainder of the growing cycle. Regardless of the lighting program, all light sockets should be utilized and burned-out bulbs replaced because dark spots affect performance. Dirty and dusty bulbs decrease the bulb life and can also reduce light levels significantly.

VENTILATION

Concern over increased brooding costs has changed house construction and brooding methods as well. For example, ventilation is becoming increasingly important as broilers are reared at higher densities and in wider, longer houses. Broiler houses are becoming more sophisticated as the trend moves from natural to power-ventilated houses.

Poultry house ventilation should supply the oxygen needs of the broilers and brooders, remove carbon dioxide and other gases such as ammonia, aid in controlling temperature and moisture, and aid in preventing potential disease problems. Drafts in houses should be prevented. Air movement in a house is dependent on temperature. It appears that the ideal temperature for broiler production after brooding is from 65 to 75°F (18.3–23.0°C).

GAS CONCENTRATIONS

Ambient air is 21% oxygen (O_2) but when open flame brooding is used oxygen content may drop to 19 or 20%. There is about 16% oxygen in expired air and respiration difficulties occur in chickens when it drops below 11%. Carbon dioxide (CO_2), a by-product of metabolism, is found normally in the air at levels below 0.004%. The ventilation system should keep carbon dioxide levels below 0.2%. Carbon monoxide (CO) occurs as the result of incomplete combustion of brooder fuel. Proper adjustment of brooders and removal of dust should prevent carbon monoxide problems. Ammonia (NH_3) is caused by the breakdown of uric acid by bacteria in the litter of poultry houses. High ammonia levels have been proven to cause increased susceptibility to Newcastle disease, as well as depressing growth rates while allowing *E. coli* organisms to proliferate. High ammonia levels (above 50 ppm) can also cause blindness in broilers. Litter moisture is the key to controlling ammonia levels since litters at 21–25% moisture levels produce little ammonia. When litter moisture exceeds 30%, ammonia production starts and increases as temperature goes up.

A general guide used by many poultrymen for determining ammonia levels is as follows:

10–15 ppm	detected by smell
25–35 ppm	eyes burn
50 ppm	watery and inflamed eyes of the broilers appear
75 ppm	broilers show discomfort, and one can observe the broilers jerking their heads.

VENTILATION FOR PARTIAL HOUSE BROODING

Since large numbers of broiler chicks are kept in a small area while maintaining high temperatures, partial house brooding with no ammonia would be difficult without power ventilation. Power ventilation should be used in partial house brooding as long as possible, even at the expense of some brooder fuel. In addition to the removal of ammonia, power ventilation removes carbon dioxide and moisture from the litter while replenishing the oxygen supply that is being depleted both by the broilers and brooders.

FAN TIMERS

It is a common practice in power-ventilated poultry houses to utilize timers on the fans rather than run them constantly. Often fans can be run as little as 5 min per hour to keep ammonia levels low. Regardless of the timer setting, the fans should be adjusted enough to control ammonia.

FAN THERMOSTATS

Fan thermostats can be utilized to turn on the fans as temperatures increase. Many companies recommend that fan thermostats be set at

3°F (1.6°C) above the desired room temperature. If more than one fan is utilized it should be set at 3°F above the setting of the first fan.

Power-ventilated houses work by exhausting air from the houses. The fans should be thermostatically controlled in hot weather to come on at 80°F (26.5°C) to prevent heat buildup in the houses.

AIR INTAKES

Air intakes, as the name implies, are the openings or controlled inlets from which air is pulled into the broiler house. Depending on the partial house brooding area, air intakes may have to open slightly. This can be determined by calculating natural air intake or by the use of a static pressure gauge. The static pressure should not exceed 0.10 in. of pressure.

Several new items of equipment are available to help keep flocks comfortable after partial house brooding. One of them is a thermostatically controlled power curtain. When activated, if temperatures change the side curtains are automatically raised or lowered day or night.

BROILER DIETS

Since the purpose of feeding broilers is to convert feedstuffs into broiler meat, rations are of major concern. Feed costs vary with the cost of ingredients but normally feed costs for broilers are 65–85% of the cost of production of the live broiler. The feed they consume is a complete ration. Broiler feeds are generally fed as crumbles for the starting feed and pellets for the remainder of the growing period.

There are several types of feeding programs used by broiler integrators. They include either the feeding of a starter and finisher or the feeding of a starter, grower, and finisher.

Table 11.6 is an example of a typical broiler feeding program.

The feed conversion in well-managed broiler houses should be in a range of 1.9–2.15 or better, depending on the nutrient density, time of year, and other factors.

CATCHING AND HAULING BROILERS

Proper catching, loading, and hauling of broilers are important in maintaining the quality achieved during the growout period. It has been shown that bruises result in 50–60% of the downgrades in broil-

TABLE 11.6. Broiler Feeding Program Example

	Starter	Grower	Finisher
Males	0–28 days	21–35 days	36 days to market
Females	0–21 days	22–42 days	43 days to market
Straight-run	0–28 days	29–35 days	36 days to market
Protein, %	21–23	19–20	16–18
Calories/pound	1425–1450	1450–1475	1475–1500
Calcium, %	0.86	0.8	0.8
Phosphorus, %	0.42	0.39	0.37
Amount per broiler[a]	1.8 lb	5 to 6	1 to 1.5

[a] Assuming 4.0 to 4.25 lb broiler.

ers. Over 30% of the bruises occur on the breast. Ninety percent of the bruises occur within 12 hr of slaughter. This indicates that most of the damage is done during catching, loading, unloading, and hanging at the processing plant.

GROWER RESPONSIBILITIES

Before catching broilers, growers should cut off and empty feed lines so the chickens can clean up the feed prior to raising the feeders. Raise the feeders 5 hr prior to loading. Raise half the waterers 30 min prior to scheduled arrival and the rest on arrival. Open curtains and doors when it is warm. Pick up all dead birds prior to loading. After loading inspect the house for damages by the loading crew and report damages within 24 hr.

CATCHING CREWS

Catching crews should drive and catch only small groups at a time to reduce smothering and bruising. Avoid kicking and throwing broilers. Always hold broilers by the shanks and feet. Place the broilers in the coops (crates) with care, never drop them. Use care in not shoving or dragging the broilers across the top of the crates.

TYPES OF BROILERS

Light broilers are grown by several integrated firms and processed from 28 to 35 days of age. The carcasses are sold whole, fresh, or frozen as Rock Cornish-type birds weighing 16–24 oz processed. Slightly heavier light broilers are also processed by some firms for further processing in their own facilities to fabricate TV dinners, frozen fried chicken, and other further processed entrees. The majority of the broilers are processed at weights of 3.5–4.5 lb live weight, which yields carcasses that can be sold whole or cut up. These broilers are processed at 42–49 days of age.

ROASTERS

A market has been developed by several companies for roasters. They are generally male broilers that have been kept until they are 10–12 weeks of age. Such birds weigh approximately 7.5–9 lb (3.5–4.0 kg) live weight. Roasters are marketed at a premium price because they are more expensive to produce than broilers.

Capons are produced by some processors on a limited basis. Capon producers have to caponize (castrate) males either surgically or chemically by implantation. Surgically caponized chickens have the widest market appeal because consumers are skeptical of chemical implants. Capons are excellent birds but modern-day roasters have most of the quality characteristics of a capon.

GROWER CONTRACTS

Broiler producers sign contracts to house and grow broilers for integrated firms. Contracts vary from firm to firm but basically all contracts guarantee the broiler grower a minimum payment per 1000 broilers grown. Many of these contracts offer a payment of $120 per 1000 broilers, as an example, plus the opportunity for a bonus based on flock performance compared to other flocks grown during that time period. Other firms pay a minimum of 3¢ per pound of broiler produced. When an average weight of 4 lb is used (4 lb × 3¢ = $0.12 × 1000 = $120) the contract is quite similar. Successful broiler producers strive to stay in the top one-third of the producers under contract. This puts them in a position to share in production efficiency bonuses and be retained as desirable producers by the integrated firm. Guaranteed grower payments are also needed by growers for use in financial planning to secure loans for broiler house construction. Recent high interest rates in the United States have limited the construction of new broiler houses.

In the future integrated firms will have several options with producers. One will be to build company-owned farms and hire managers to produce broilers for the firm. This method has advantages and disadvantages and requires large capital expenditures. A more feasible short-term solution would be to guarantee producers a minimum contract with an interest allowance clause that could be tied to floating interest rates. Under these conditions integrators could be assured that adequate housing would be available and that during times of high loan interest rates the burden of the interest increase could be subsidized and absorbed by the firm as a production cost, which will eventually be passed to the consumer as a slightly higher price broiler.

Regardless of the system used, in the future broilers will continue to supply an ever increasing portion of our meat available for consumption as additional efficiencies are achieved.

REFERENCES

Benson, V. W., and Witzig, T. J. 1977. The Chicken Broiler Industry. Agricultural Economic Report 381, USDA, Washington, DC.

Brooks, R. C. 1980. The U.S. Broiler Industry. Tar Heel Economist, Agricultural Extension Service, N.C. State University, Raleigh, NC.

Brown, R. H. 1979. The Sunday Dinner Becomes Everyday Fare, Chapter 9, The Broiler Industry. Feedstuffs, October 29, 1979.

Cobb, Incorporated. 1982. Cobb Broiler Manual. Subsidiary of the Upjohn Company, Concord, MA.

Crowley, J. 1977. People on the Farm: Broiler Growers. USDA, Washington, DC.

Harris, J. R. 1964. Minimum Standards for Profitable Broiler Production. The North Carolina State Poultry Extension Service, Raleigh, NC.

Mills, W. C., Martin, G. A., and Ward, J. B. 1975. North Carolina Broiler Breeder Program. Circular 482, The North Carolina Agricultural Extension Service, Raleigh, NC.

North, M. O. 1984. Commercial Chicken Production Manual, 3rd Edition. AVI Publishing Company, Westport, CT.

USDA. 1983. Agricultural Statistics. U.S. Government Printing Office, Washington, DC.

USDA. 1984. Foreign Agricultural Circular: Dairy, Livestock and Poultry Division. FL&P-184, Washington, DC.

USDA. 1987. Statistical Reporting Service, POU 3-1 (87).

12

Commercial Egg Production

Mass production of chicken eggs has become a highly efficient, competitive enterprise. Several farms in the United States now have over two million birds on a single farm. The following example illustrates the complexity of these operations. To set up and operate a 300,000 layer farm requires an investment of over two and one-half million dollars. The hens consume about 35 tons of feed a day and 16,000 gal of water. They produce 75,000 lb of manure and 600 cases of eggs daily. Equipment consists of 4–5 miles of watering troughs or cups, 14 miles of feeders, 14 miles of egg belts, and several hundred fans and electric motors. Only one person is required to handle 100,000 birds.

The pullets will begin laying eggs at 20–22 weeks of age and once laying will eventually peak at 85–93% production. Each bird can be expected to lay 270 quality eggs by the age of 75 weeks. About 4 lb (1.8 kg) of feed for each dozen eggs, which will weigh about 1.5 lb (0.7 kg) per dozen, is needed to produce these eggs.

The growth and automation of the commercial egg manufacturing industry have developed faster and progressed further than any other type of livestock production during a period of only about 40 years. Table 12.1 illustrates the progress made in reducing space requirements and body weights during this period.

TABLE 12.1. Changes in Space Requirements and Body Weight for Laying Hens[a]

Year	Type of housing	Space per bird		Body weight	
		ft² (m²)	ft³	lb/bird	kg/bird
1926	Floor	3	32	4	1.8
1946	Floor	3	24	4.5	2.0
1966–1967	Floor	1	8	4	1.8
1966–1967	Cage	0.5	4	3.9	1.8
1970	Cage	0.3	2.5	3.9	1.8

[a] Reconstructed from data of Patrick and Schaible (1980).

In the United States the commercial egg industry utilized many dual-purpose breeds of layers prior to the 1950s. Breeds of dual purpose chickens such as the New Hampshires, Rhode Island Reds, and Barred Plymouth Rocks were more popular than the leghorn-type layers utilized today. The dual-purpose breeds of chickens were popular because the cockerels could be sold as meat-type birds and the majority of consumers had been accustomed to brown-shelled eggs and preferred them. As the commercial egg industry developed and become more competitive and cost conscious, the brown egg layer became less prominent. The broiler industry was developing and the dual-purpose cockerels did not grow as rapidly as the commercial broiler crosses. The white egg laying commercial leghorn strains available to the industry had a high rate of lay and a smaller body size that allowed leghorn-type stocks to produce a dozen eggs on less feed. The uniformly white eggs make a more uniform product to package for consumers. Today over 90% of the commercial layers are leghorn type. Unless a premium can be obtained for brown-shelled eggs, production efficiencies favor the leghorn stocks. Some areas of the United States, such as the northeast, still have strong brown-shelled egg markets.

With varying amounts of success, geneticists have recently introduced the dwarf gene into some leghorn stocks to reduce body size even more in an attempt to lower feed costs by maintaining less weight in the layer.

Originally, laying hens were kept in houses with floors of dirt, wood, or concrete covered with litter such as straw, shavings, peanut hulls, ground corn cobs, sugar cane, or other types of litter. Some operators have used wire floors or wood slats successfully. The houses were usually equipped with roosts, waterers, and feeders, which were hand cleaned and filled by the operator, nests to collect the eggs, lights, and a natural ventilation system controlled by louvres or windows. As the industry progressed, mechanical feeders and waterers were installed, roosts were eliminated, and automatic timers for light and ventilating fans were used along with insulation. Gradually, over a 40-year period, houses with almost complete environmental control were developed and laying hens began to be kept in individual cages. Gradually two, three, and even four hens were kept in the same cage. The use of cages made increased bird housing densities and the use of more automation possible, which resulted in the housing cost per bird in such operations being less than in traditional floor houses. Finally, equipment was designed in such a manner that as soon as an egg was laid it rolled down to a conveyor belt which moved the eggs into a processing room in which they were automatically counted, washed, and graded for size and quality.

Attempts to rear pullets and keep laying hens in cages started on a small scale during the 1930s. After a number of attempts in full-scale commercial operation, the process was discontinued except for brooding chickens in batteries and for a few single cage operations in California. The latest trend of keeping laying hens in cages started after World War II. Since that time the percentage of laying hens kept in cages on farms has increased to the point at which over 91% of all commercial laying hens are now kept in cages. Most breeding flocks are still reared in floor pens.

To obtain laying hens producers used to start and rear day-old chicks on their own farm in brooder houses. Generally, the males were sold as frying chickens. When the birds were able to survive outside in the spring, they were moved to fenced grass pastures with simple range houses for protection against rain and predators. In the fall, the ready-to-lay pullets were transferred to the laying house to replace older hens going out of production. Gradually, to improve efficiency and reduce production costs, some hatcheries and producers began to specialize in the production of ready-to-lay pullets, which they sold to laying cage operators for egg production. At the same time pullet producers found that pullets could be started and raised in total confinement in the same house until ready for delivery to cage operators more efficiently than on range and with no decline in quality. Successful systems have now been developed in which day-old pullets are started and reared to laying age in special pullet-rearing cages. Many pullets are still started on the floor with conventional brooding systems, but when they reach 6–8 weeks of age, they are transferred to cages. New cage pullet-rearing systems are constantly being tested under commercial conditions so that additional refinements and improvements are continuing.

The trend to rearing replacement pullets in cages has been growing rapidly. In 1972, it was estimated that about 21% of all replacement pullets were reared in cages; by 1980, over 40% were being reared in cages. At the present time innovation, experimentation, development, and improvements in rearing replacement pullets and housing laying hens in cages are continuing with rapid adoption of the more successful ideas by industry.

STARTING PULLETS IN CAGES

Started pullet growers attempt to produce at minimum expense healthy, well-fleshed, and feathered 18 to 20-week-old, ready-to-lay pullets with maximum egg production potential. Table 12.2 lists the items and costs, less feed, for producing a pullet in 1984.

Patrick and Schaible (1980) have listed the following advantages to egg producers in buying started pullets:

1. Better use of time may be obtained by becoming a specialist in egg production.
2. Better use of buildings and equipment may be made (other than for brooding and raising chicks).
3. No money will be tied up for long periods and money will be saved which would otherwise be invested in brooding facilities.
4. The risks of starting and growing chickens are cut.
5. The disease hazard is reduced when all birds are marketed at once so no birds are left on the premises to infect the next flock of pullets.
6. Better pullets may often be purchased at a cost less than the poultryman can raise them himself.

TABLE 12.2.   Cost of 20-Week-Old Replacement Pullets[a]

Cost item	1964	1968	1984
Feed	$0.625	$0.634	$1.257[b]
Chicks	0.331	0.304	0.515
Brooder fuel	0.006	0.012	0.038
Utilities	—	0.006	0.040
Medicine	0.013	0.024	0.006
Vaccine	0.058	0.042	0.060
Labor	0.085	0.109	0.281
Management	—	0.009	0.039
Taxes	—	0.013	0.013
Trucking	—	—	0.045
Water	—	—	0.006
Housing and equipment	0.046	0.076	0.108
Interest	0.047	0.045	0.135
Land	—	—	0.011
Miscellaneous	0.004	0.014	0.036
Total	$1.21	$1.29	$2.59

[a] Source: Bell (1986).
[b] Feed costs were higher than average in 1984.

7. The culls in the laying house may be reduced because of the purchase of Grade A pullets.

Disadvantages include the following:

1. Disease hazards exist.
2. Costs may be increased (the grower has to make a profit).
3. It is not always possible to get started pullets of the strain desired.
4. The feeding program or management cannot be controlled.

BROODING

Chicks can be vaccinated, dubbed, and debeaked at the hatchery if desired. Vaccinations should be in line with the prevalent diseases in the area and according to the local veterinarian or hatcheryman's recommendations. When a second debeaking is required, it should be done before the chicks are 16 weeks old. Some growers also have the chicks front three toes clipped at the hatchery or before they are placed in the cage brooders.

The same general rules of sanitizing houses and equipment and preparation for the chicks apply for pullets reared in cages as for those reared on the floor. These include placing starter paper on the cage floors before the chicks arrive and manually filling the feeder troughs even with the edge. Waterers should be adjusted to the lowest height and filled full. The rooms should be heated to 85°F (30°C) at least a day before the chicks arrive, lighted continuously with a light intensity of 1–2 ft candles (10–20 lux) after the chicks arrive, and ventilators should be adjusted so that there is a change of about 10 ft³/min of air for each 100 chicks. Several hours after the chicks have been placed in the cages, feed is put in the cages on filler flats to help all chicks eat.

Chicks should be observed constantly during their first few hours in cages with special attention being given to whether they are comfortable and eating and drinking. If the chicks are cold they will huddle together and if they are too warm, they will spread out, pant, and keep very quiet.

As soon as the level of feed in the hoppers has been consumed to the normal operating level of the feeders, the feeders should be turned on and off several times before letting them run continuously. This practice gives the chicks a chance to get used to them. Other hints to help chicks adapt to automatic feeders include a full circuit manual operation twice a day to fill the troughs as full as possible.

After 3 or 4 days, light intensity should be gradually reduced to 0.5–1 door candle (0.5–1 lux) at feed level. After the first week feeders with adjustable speeds can be set to operate automatically at about 20 ft/min and, after the third week, the feed level should be reduced to prevent waste and the speed of the feeders increased to 40 ft/min.

As chicks grow, water supplies should be increased to provide easy access and to reduce spilling. If cups are used, water pressure should be increased gradually.

After about 1 week, the room temperature should be reduced gradually about 3°–5° a week until a temperature of 70°F (21°C) is reached. Some producers prefer to reduce the temperature down to 60°F (16°C) to stimulate more rapid feathering, but such a practice also increases feed consumption 5–10%. Ventilation should also be increased gradually to 75 cfm per 100 pullets by 5 weeks of age.

GROWING PULLETS IN CAGES

Some pullet chicks are started and reared in the same cage until they are transferred to laying cages. This type of system is called the brood-grow cage system. Others are started and reared in cages or on the floor, usually to about 6–8 weeks of age and then transferred to growing cages (Fig. 12.1).

Feeding and lighting are manipulated to have the birds mature physically and sexually so they will begin producing small to medium eggs about the time they are transferred to laying cages. If the birds gain weight too fast, feed intake should be reduced; if they do not gain weight fast enough, feed intake should be stimulated. Since the activity of caged birds is more restricted than floor birds, feeding practices are slightly different. Major breeders supply a growth curve chart for their strain of birds to help producers stay on target by weighing samples of pullets at different ages and adjusting the feeding program as needed.

Pullets frequently mature sexually faster than they mature physically. As a result they tend to lay small eggs if they start producing before they are physically mature. The physical–sexual maturity balance is usually achieved by varying the feeding schedule and the amounts and intensity of lighting. Fall-hatched pullets reared under natural light always tend to lay earlier than spring-hatched pullets because they develop under increasing day length.

A number of benefits accrue from keeping growing pullets on a restricted feeding schedule (Fig. 12.2). Restricted birds generally require 5–10 days longer to reach sexual maturity. As a result, they lay fewer small eggs when they start to lay. Generally, birds reared on such programs also have lower mortality and consume less feed during their growing and laying periods because the body weights of such

FIG. 12.1.   The trend to rearing replacement pullets in cages has been growing rapidly. Shown here are some of the 40,000 pullets in a 420-ft automated chicken house in South Carolina. (Source: USDA.)

birds are slightly less than their full-fed counterparts, since they have less body fat and are generally more heat tolerant. In addition, birds on restricted feeding schedules and hatched from October through February generally lay more eggs. Producers generally do not practice feed restraint of pullets in the United States to the degree that it is practiced in other parts of the world in which feed is more costly.

A number of other factors must be considered when feeding growing pullets. One of the more important ones is the cost and amounts of feed consumed. In cold weather the poultry producers must decide whether to keep the birds warm by heating the house or by supplying extra feed. Currently, producers are building better insulated housing and managing them to conserve both feed and energy as much as possible.

Crowding of pullets in cages is also a matter of concern. It not only

FIG. 12.2.   The use of automatic feeders is now standard practice in the industry. (Source: USDA.)

reduces body weight at maturity but also reduces future egg production of pullets. Overcrowding of pullets in cages has little merit from an economic standpoint or from a management viewpoint since the stress of overcrowding causes less uniformity and more cull pullets.

MANAGING LAYING HENS IN CAGES

The large increases in production efficiency and the decrease in housing and labor costs per bird when hens are kept in cages are more than enough to compensate for the disadvantages. Progress in improving and modifying cage layer systems is still continuing in the industry with many combinations. Figure 12.1 shows some typical cage arrangements currently in use.

Feeding can be done manually in which case an operator fills a continuous trough from a moving cart. The use of automatic feeders is now standard practice in the industry. With automatic feeders, feed is moved from a central feed storage bin outside the building to the troughs by means of a moving chain, an auger-type conveyor, or a cable with flexible disks attached.

Water can be supplied from troughs with slowly running water, or individual cups or nipples. Individual cups or nipples are used entirely in new installations because they conserve water and have fewer birds drinking from the same waterer, so the potential for disease spread is lessened.

Eggs can be gathered manually by using the same cart used for feeding or automatically on moving belts, which not only convey them to a central point, but count them as well. The automatic gatherer can be attached to a washer, grader, candler, and packer if desired. The newer and larger cage layer facilities all utilize automatic egg gatherers because of the savings in labor.

Manure is managed by permitting droppings to come directly under cages, drop into pits or lagoons, or by having scrapers which scrape the droppings to the end of the building (Figs. 12.3 and 12.4). Insect infestations, particularly from flies, khapra, and darkling beetles are particular problems in cage systems. Proper control of manure, particularly by keeping it dry, and frequent and proper use of insecticides are the best control measures.

A number of management combinations to achieve maximum number of eggs per hen and maximum return per dollar invested have been tried under a wide variety of conditions. Several general principles have evolved from these experiments.

In general, each additional bird added to a cage will cause a decrease in egg production and an increase in mortality. Birds kept in cages with 6 in. (15 cm) of feeder space are more profitable than those with less feeder space. In general, when cages are placed lengthwise parallel to the aisle compared to the width fronting on the aisle (reverse cages), net income per hen is increased.

Table 12.3 shows the performance and economic results of housing 1,

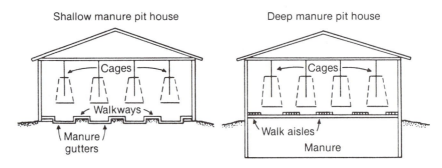

FIG. 12.3. Layer cage manure disposal systems.

FIG. 12.4. A fast-moving stream of water moves chicken manure quickly from a laying house to a lagoon. (Source: USDA.)

2, 3, or 4 hens in 12 × 18 in. cages (30 × 45 cm). The maximum number of eggs produced per hen is not always the most accurate measure of profit.

INDUCED MOLTING

Since the cost of a pullet is one of the largest costs in an egg operation, poultrymen must extend the productive laying life of their birds as long as profitable. To ensure satisfactory production, birds are frequently molted. A typical way to bring on molting, which is usually followed by cessation of egg production, is to restrict light and remove feed and water for several days. The California Extension Service has the following recommendations for force molting.

TABLE 12.3. Performance and Economic Results of 1, 2, 3, or 4 Hens per 12 × 18 in. Cage ($100,000 Invested in 4000 Cages) [a, b]

Performance results	Hens per cage			
	1	2	3	4
Hen–day production (%)	73	69	69	65
Eggs per hen housed	284	267	264	241
Mortality (%)	4.4	5.7	6.7	15.7
lb feed/dozen	3.8	3.9	3.9	4.1
Egg income minus feed cost	$3.99	$3.65	$3.61	$3.08
Returns per cage				
Income				
Eggs	$10.27	$19.27	$28.84	$35.27
Fowl	0.33	0.66	0.98	1.18
Total	$10.60	$19.93	$29.82	$36.45
Costs				
Pullets	2.00	4.00	6.00	8.00
Feed	6.29	12.00	17.94	22.88
Total	8.29	16.00	23.94	30.88
Income minus cost	$2.31	$3.93	$5.88	$5.57
Investment vs. returns				
Investment bird	$25.00	$12.50	$8.33	$6.25
Income/house	$9,240	$15,720	$23,520	$22,280
Return on initial investment	9.2%	15.7%	23.5%	22.8%

[a] Source: Bell (1986).
[b] Forty-five cents per dozen for large eggs and $7.00 per 100 lb for feed.

For a rapid molt (return to 50% production in less than 6 weeks):

Step 1. Turn off artificial lighting in open housing or reduce to 8 hr in enclosed housing on day 1.
Step 2. Remove all feed for 10 days. Do not remove water.
Step 3. Provide shell during the 10 days.
Step 4. Following the 10 days of no feed, full feed a regular laying ration and turn lights back to the normal program.

For a normal molting (return to 50% production in 6–8 weeks):

Step 1. Turn off artificial lighting in open housing or reduce to 8 hr in enclosed housing on day 1.
Step 2. Remove all feed for 10 days. Do not remove water.
Step 3. Shell feeding is optional.
Step 4. Starting on the eleventh day, full-feed cracked grain for 2–3 weeks.
Step 5. At the end of the grain feeding period, feed a normal laying ration and turn the lights back on.

For slow molting (return to 50% production in more than 8 weeks):

Step 1.   Turn off artificial lighting in open housing or reduce to 8 hr in enclosed housing on day 1.
Step 2.   Remove all feed for 10 days. Do not remove water.
Step 3.   Do not feed shell.
Step 4.   Starting on the eleventh day, full-feed cracked grain for a period of 4–5 weeks.
Step 5.   When ready to bring the flock back into production, feed a normal laying ration and turn the lights back on.

Molting of layers requires good management and attention to detail. Proper light control is important and exposure of hens to increasing light must be avoided. Molting cannot make a good flock out of a mediocre one so only high-performing flocks should be considered. Strains of birds respond differently to induced molting so be flexible in the molt procedure. Alternative molting programs to feed restriction have been developed, such as low dietary salt or high dietary zinc that can produce satisfactory results. In molting hens remember that there are no substitutes for proper flock supervision, body weight monitoring, feed management, and light control.

REFERENCES

Bell, D. O. 1986. Personal communication. California Agriculture Extension Service, Riverside, CA.

Brake, J. T., and Carey, J. B. 1983. Induced Molting of Commercial Layers. Poultry Science & Technology Guide No. 10, NCSU at Raleigh Extension Poultry Science.

Carter, T. A., and Vocke, G. F. 1978. Contract Commercial Egg Production. PS&T Guide No. 12, NCSU at Raleigh Extension Poultry Science.

Morris, T. B., Martin, G. A., and Mills, W. C., Jr. 1972. Egg Size and Profit. PS&T Guide No. 14, NCSU at Raleigh Extension Poultry Science.

North, M. O. 1984. Commercial Chicken Production Manual, 3rd Edition. AVI Publishing Company, Westport, CT.

Patrick, H., and Schaible, P. J. 1980. Poultry Feeds and Nutrition, 2nd Edition. AVI Publishing Company, Westport, CT.

13

Turkey Production

ORIGIN OF THE TURKEY

The turkey is the only major meat animal to come from North America. Although Indians domesticated turkeys in Mexico, the southwestern United States, North Carolina, and Virginia as early as A.D. 400, the early civilized world was not aware of this bird until after the discovery of North America. The scientific name of the turkey is *Meleagris gallopavo*, *Meleagris* being the Latin name for guinea fowl and *gallopavo* from gallus meaning cock and from pavo meaning peafowl. Turkeys belong to the genus *Gallus* of the family Phasianidae. The origin of the word turkey has been lost, but the turkey's name may have come from an aboriginal Indian word "furkee" or from its own call note: turk, turk (Fig. 13.1).

Spanish explorers took Mexican wild turkeys domesticated by the Aztecs home to Europe. The turkey then spread rapidly throughout Europe and was introduced in England between 1524 and 1541. The colonists then returned these domesticated turkeys to the colonies. Our present domesticated turkeys are descendants of the Mexican subspecies, *Meleagris gallopavo gallopavo* and the eastern wild turkey *Meleagris gallopavo silvestris*.

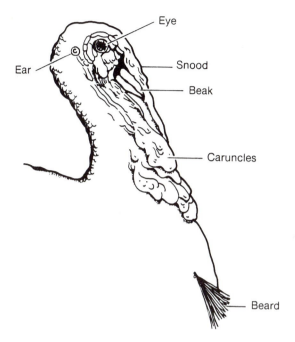

FIG. 13.1.   Anatomical features of the turkey.

COLOR VARIETIES

The black color variety of turkey evolved from the selection of darker birds in the population, which was a common practice of southern Europeans breeding poultry. Black turkeys were particularly popular in Spain, France, and Italy. Even now there is a market for black and bronze birds in these countries. In England, the bronze-colored birds became the predominant variety.

As far as can be determined, our domestic turkeys started from the black and bronze turkeys brought from Europe by the colonists. Later in Rhode Island in the Narragansett Bay area, domestic stocks were crossed with eastern wild turkeys to develop the Narragansett turkey, one of the forerunners of our present day varieties and crosses.

Later, around 1830 in this same area, Narragansett turkeys were crossed again with the eastern wild turkey to develop the bronze variety. About the same time, a white turkey known as the Austrian White was developed in Europe. After looking at the genetic make-up for color, it is generally thought that the White Holland evolved as a sport from the bronze.

The Bourbon Red, developed in Bourbon County, Kentucky, is probably the most attractive variety. It is reddish-brown in color and has white primary, secondary, and main tail feathers. Other varieties that have been developed include the slate, buff, and royal palm turkeys.

The USDA started to develop a small turkey called the Beltsville Small White in 1934. This variety was introduced in 1941 and is noted for its small size and thick, meaty breast. In today's industry this type of turkey is no longer utilized.

When turkeys moved westward in the United States, some growers began to breed turkeys in the northwest. Later when dressed turkeys began to be shown at poultry shows breeders began to emphasize body conformation more than feathering.

During this period, Jess Throessel of British Columbia imported some Sheffield strain Bronze turkeys from England. The large bronze toms from this strain were used by breeders to mate to females selected for body fleshing. Because of their meaty breasts, the offspring, called the Broad Breasted Bronze, were in great demand.

By 1955 dressed, ready-to-cook turkey carcasses began to replace the live turkeys in sales. Since turkeys have a white skin, the dark pinfeathers and melanin pigments left in the follicles when the growing feathers were removed from dark feathered varieties produced an unattractive carcass. This led to the development of a broad white turkey, accomplished by crossing the Broad Breasted Bronze with a white feathered variety, then backcrossing again to the Broad Breasted Bronze, which produced some bronze-colored and some white-feathered birds. By selecting and breeding only the white offspring, white-feathered, broad breasted varieties were developed. Gradually the size and conformation of the white turkey approached that of the Broad Breasted Bronze. Our present white turkey foundation stocks are essentially Broad Breasted Bronze turkeys masquerading in white feathers.

TURKEY GENETICS

The diploid chromosome number in turkeys is 82 compared to 78 in the chicken. Because the sex chromosome has only one color gene, color in turkeys is controlled almost entirely by the difference of one gene; some color patterns such as Narragansett are sex linked. All the genes present in turkeys are in either the dominant or recessive trait, never heterozygous. The development of various color patterns as well as the genetic components contributing to them are listed in Table 13.1. This table shows that N or n and the Y on the chromosome indicate the sex is female. Examples are NY or nY. All others would be male. By using

TABLE 13.1. Genetic Formulas for Turkeys Pure for the Color Pattern Shown[a]

Color	Male	Female
Bronze	WWddNNbbRR	WWddNYbbRR
Bourbon Red	WWddNNbbrr	WWddNYbbrr
Narragansett	WWddnnbbRR	WWddnYbbRR
Black	WWddNNBBRR	WWddNYBBRR
Buff	WWddNNBBrr	WWddNYBBrr
Slate	WWDDNNbbRR	WWDDNYbbRR
White	ww (with any above combination)	ww (with any above combination)

[a] W, Nonwhite; w, white; N, Narragansett; n, non-Narragansett; D, slate; d, recessive slate; b, nonblack; B, black; R, red.

this table one can see how the current varieties of turkeys evolved from wild turkeys.

Our present turkey-breeding stock was developed much like broiler-breeding stock. Many breeders used phenotypic selection to improve physical traits. In some instances pedigree systems with trapnesting were used to develop pure lines with emphasis on reproductive traits. Live and dressed turkey shows influenced breeders to select for meat characteristics and consumers demanded more meaty breasts on the turkeys they purchased. More recently the hatchery phase of the industry asked for improvement in reproductive traits, particularly egg number and hatchability.

Only a few turkey breeders remain today. One of the most successful was George Nicholas, who founded Nicholas Turkey Breeding Farms of California. The Nicholas strain was developed from several other strains and then improved by using a full pedigree program. The eventual result is a turkey strain with both good meat characteristics and good reproductive traits. The original Nicholas turkeys were bronze feathered but were changed by breeding to a white-feathered strain in the 1960s. A two-way cross is used on the female side while the pure line has been retained on the male side resulting in a commercial market turkey that is a three-way cross.

Other suppliers of turkey-breeding stock include British United Turkeys in England and Hybrid Turkeys in Ontario, Canada. British United Turkey provides most of the breeding stock in Europe. Scottish interests started the original breeding program by using a full pedigree program with limited trapnesting.

Hybrid Turkeys in Ontario, Canada has developed some good strains of medium and large turkeys. They sell turkeys in both Canada and the United States and they are the only other breeder doing pedigree breeding at this time.

TURKEY BREEDER FLOCKS

In turkey breeder production feed cost is about another 70% of the cost of production. The large cost in turkey production is poult cost because egg production and hatchability are poorer than for chickens. For these reasons, considerable research and progress have been made in increasing egg production and improving fertility and hatchability. Breeding and management, in particular lighting and artificial insemination, have helped to increase the yield of healthy poults.

TURKEY BREEDERS

Integrated turkey producers and independent hatcheries today secure the male and female lines directly from a primary breeder, such as Nicholas, as hatching eggs. The eggs are hatched and the poults are sexed at the hatchery. In the male line tom poults are reared for selection as potential breeders and the male line hens are slaughtered at market age. The reverse happens in the female line; the female line toms are slaughtered. Generally, about one tom poult is started for every five hen poults. Potential breeders are selected at about 16 weeks of age for the hens and 16 to 18 weeks for the toms. Birds with crooked beaks, roached backs, pendulous crops, crooked toes, bowed legs, swollen hocks, blind eyes, and crooked keels as well as birds below a certain weight limit for both toms and hens are undesirable as breeders and are culled from the flock. Toms are also selected on the basis of conformation. Breeders are also blood tested at this time for *Mycoplasma meleagridis, Mycoplasma synoviae,* and pullorum. A vaccination program suitable for the area is also carried out to protect the investment in breeders.

Turkeys reach sexual maturity at an older age than chickens. For this reason they are generally not light stimulated until they are between 29 and 32 weeks of age. A second culling usually occurs when the breeders are moved to breeder houses. Since breeder hens and toms are housed separately, care should be taken to ensure that the toms are adequately light stimulated to produce viable spermatozoa when the hens are ready to be inseminated. A common lighting schedule for toms is 12 hr per day at an intensity of 5 ft candle (50 lux) at bird's eye level beginning at 18–20 weeks of age. This regime is followed throughout the breeding season in many instances. A light program of this type requires total environmental control.

TURKEY BREEDER TOMS

Breeder toms (Fig. 13.2) are housed in separate buildings from the hens but on the same farm. Toms are usually housed in small groups of 10–15 birds per pen to reduce injuries from fighting and are normally given roughly 5 to 10 ft^2 (0.5–1.0 m^2) of floor space. Some breeders restrict the feed intake of breeder toms to prevent them from becoming overweight. Feed restriction also improves the overall quality of semen as well as the duration of collection of quality semen. A sexually mature breeder tom has a value of approximately $50. Each tom has the potential to sire through artificial insemination over 1000 poults through the 15–20 hens that can be inseminated with his semen.

TURKEY BREEDER HENS

Turkey hens are managed to produce their first eggs at about 32 to 36 weeks of age. The production cycle of a turkey hen lasts roughly

FIG. 13.2. A modern turkey breeder tom.

6 months. During that time, a hen should produce 75–90 eggs depending on the time of year and other related factors. Since turkey hens lay relatively few eggs, the value of an individual egg is high compared to chicken breeder eggs. A sexually mature turkey hen has a value of roughly $25 and a fertile turkey egg probably has a value of around $0.55–0.65.

Turkey hens are generally housed in total confinement or semiconfinement houses (Fig. 13.3). The trend tends to be toward total confinement in better quality houses to permit optimum control of lights for off-season egg production. Turkey hens are normally given 3–7 ft^2 (0.3–0.5 m^2) of floor space per hen. Both hens and toms are kept on litter on concrete, asphalt, or dirt floors.

Only limited success has been achieved rearing turkey hens in cages because of foot and leg problems associated with standing in the cage. No commercial breeder hens are managed in cages in the United States. Breeder pens normally have nests that have semi- or full-trap fronts. This prevents egg breakage from multiple use of the same nest and helps prevent stimulation of prolactin which causes broodiness. Eggs are gathered frequently, 6–10 times a day. Hens are pushed from the nest at each gathering. This assists in broody control.

FIG. 13.3.   Turkey breeder hens confined in the house. Note the trap nests.

ARTIFICIAL INSEMINATION

Artificial insemination is used entirely in commercial turkey production. High fertility cannot be achieved by using natural matings. Since breeders placed so much emphasis on breast length and width in their selection programs, males cannot mate satisfactorily. Heavy breasts on toms tend to move their center of gravity forward so that when toms mount hens to mate either the males and females are no longer matched for mating or the toms topple off the hens before copulation. Both of these situations tend to affect the libido of the toms and fertility suffers. Many hens are also injured and stressed by unsuccessful attempts at mating.

Artificial insemination can be accomplished by training lightstimulated toms for ejaculation and semen collection. The toms are held and massaged on the back to stimulate treading, then the tail head is forced toward the tom's head over the back while pressure is applied to the sides of the cloaca and genital ridge area. A tom can be ejaculated about twice a week and will yield about 0.25 ml of semen per ejaculate. At present, semen can only be stored a short time. It is usually used within 1 hr of collection.

Semen from several toms is usually pooled and a commercial semen extender, used at various dose levels, is added before inseminating hens. For insemination, a turkey hen is held so that pressure is applied to the abdomen which everts the functional left oviduct to the surface of the cloaca. A semen straw containing 0.025 ml of semen or about 100 million spermatozoa is used per insemination. The semen straw is gently placed up to 2 in. (5.0 cm) into the oviduct and then the semen is gently evacuated from the straw by air pressure. Disposable straws are used as an aid in preventing disease. Commercially, turkey hens are inseminated prior to laying their first egg and then weekly as long as hatching eggs are collected. At one time, inseminating crews went from farm to farm to inseminate turkeys. Now on many farms, a crew on the breeder farm inseminates and collects semen to prevent the risk of introducing disease in the flock.

HATCHING EGGS

The incubation period of turkey eggs is 28 days. The poults are normally removed from the hatchers on the twenty-ninth day.

Hatching eggs are collected frequently at the farm, cooled to 55–65°F (13–18°C), and held for delivery to the hatchery two or three times each week where they are washed and sanitized. In most hatcheries, the eggs are treated with an antibiotic by using a pressurized container under a partial vacuum before being placed in the incubator.

HATCHERY SERVICES

Young turkeys, called poults (Fig. 13.4), are removed from the hatchers and then vent sexed so the toms and hens can be reared separately. The poults' beaks are then blunted to prevent cannibalism and the snood is removed to prevent swelling and injuries from fighting. Toes are removed just behind the nails to prevent the birds from scratching each other when they become frightened during loading and also from trampling each other. Sometimes poults are given an injection of an antibiotic subcutaneously behind the neck to prevent bacterial infections.

A newly hatched poult has a value of about $1.00. However, in recent years, because of the demand for further processed turkeys, hen poults have been discounted and toms have been in short supply at higher prices.

FIG. 13.4.   A newly hatched turkey poult. (Source: USDA.)

BROODING

Turkeys are usually started in brooder houses similar to those used for chicken broilers.

Supplemental heat may be needed for 2–6 weeks depending on the part of the country and time of year.

After 6–8 weeks, the poults are transferred to open ranges or houses in which they are reared in total confinement. A typical brooder house is 40 ft wide (13 m) and 250–400 ft long (80–130 m) with a 15 ft high (5 m) metal roof insulated with 3.5 in. (9.0 cm) of Fiberglas insulation, a dirt floor, often of packed clay or impregnated with diesel oil, and plastic or solid curtains on the sidewalls, depending on the climate. Most houses are equipped with automatic feeders, drinking fountains, and canopy-type brooders. A recent trend has been to use only part of the house for brooding the first 3 weeks to save brooder fuel and then to open the entire brooder house to the poults for the last 3–5 weeks. Many integrated firms advise growers to build houses in clusters of one brooder house and two growout houses. After brooding, the poults are split into the two grower houses which are less costly to maintain so the next group of poults can be started in the brooder house.

The North Central Regional Extension Services have prepared the following recommendations for brooding poults.

FLOOR BROODING

Preparing for Poults

Thoroughly clean, disinfect, and test the brooder house and equipment well before the poults arrive. Clean out old litter and accumulated manure and remove dust from sidewalls and ceilings. The entire interior should also be thoroughly washed down. Pressurized washers are especially useful for cleaning walls, ceilings, and hard-to-reach areas. Cleaning solutions may speed cleaning but they are not essential. Water under high pressure can do a good job.

Follow the cleanout with a good disinfectant spray. Disinfectants will not penetrate accumulated dirt or manure, so they are no substitute for a thorough cleaning. Rinse the house down again if recommended on the disinfectant label and allow the house to dry completely. Airing the house for several days also helps control diseases.

Put down 2–4 in. (5–10 cm) of litter; have all equipment in place and make sure it is operating properly 2 or 3 days before the poults arrive. Commercial turkey producers reuse litter in many areas of the country as is done with broilers. Early preparation allows time to make repairs and adjustments. Brooders should be operated at the proper temperature at least 24 hr before poults are delivered to warm the house and so thermostats can be adjusted. Feed and water should be available so poults can start eating and drinking immediately after arriving.

Poults should be put under the brooders as quickly and quietly as possible. Sudden loud noises might cause them to crowd. The poults should be checked quietly and frequently, particularly just before and after dark.

FLOOR COVERING

Litter materials should be clean, dry, absorbent, and dust free. Shavings, cane pulp, rice hulls, ground corn cobs, chopped straw, and other commercial litters may be used, depending on cost and availability. Ground corn cobs, however, may harbor molds or support mold growth. Any organic litter will support mold growth if it remains in the pen after getting wet.

Excessively fine litter materials that poults are likely to consume should not be used. Litter consumption may increase mortality because of crop compaction. Litter can be covered with burlap or rough (crinkled) paper for the first few days to keep the poults from eating litter; it should not be covered with slick-surfaced materials since these may cause serious leg problems. Newspapers, for example, are too smooth to provide proper footing for poults. This paper or cloth should be removed when the poults are 4 to 6 days old.

Wet or caked litter should be removed and new litter added to maintain a depth of 2–4 in. (5–10 cm). Stirring wet litter helps maintain its quality as poults get older.

Slat floors, sometimes used to brood and rear poults, should be covered with rough paper or cloth for starting poults. How long this covering remains over the slats varies with the type of slats.

BROODERS

Inexperienced turkey growers should not brood more than 250 to 300 poults under each hover although some experienced growers may brood more. Allow at least 12 in.^{2} (30 cm^2) of floor space per poult under full hovers that confine heat. Radiant brooders have smaller hovers but extend the heat area outward. Follow the manufacturer's recommendations on capacity for poults.

Popular brooding systems are heated by oil, natural gas, bottled gas, or electricity. Choose a system which relies on the least-costly fuel available. There is more risk of fire with open-flame brooders and they must be kept in good operating condition. Catalytic and electric brooders present less fire hazard. The risk of fire with hot water systems is reduced since the furnace unit is located outside the brooding area.

Experienced turkey growers can select the optimum brooding temperatures by observing poults. They will huddle together under the hover when the temperature is too low and will move out against the brooder guard ring when the temperature is too warm. Poults that consistently stay on one side of the brooding area are trying to escape a drafty area. Comfortable poults will spread uniformly under and around the edge of the brooder. Poults are the most accurate "thermometer," but growers need some experience to interpret the reactions of poults.

A thermometer should be placed at the edge of the hover, about 3 in. (7.5 cm) above the litter. The brooding temperature should be 95°F (35°C) for the first week and should then be lowered by 5°F (3°C) each week thereafter until it reaches 65–70°F (18–21°C). Recent research has shown that these temperatures can be lowered about 5°F (3°C), but doing so reduces margins for error and requires extra management.

Space heaters may be used to help raise room temperatures and reduce the brooder stove output. Temperatures near the floor of the room outside the brooding area should be maintained at 70–75°F (21–24°C) for the first week and then gradually reduced to 55–60°F (13–16°C). Poults require more litter and consume more feed at lower room temperatures.

BROODER GUARDS

Brooder guards are barriers placed 3–5 ft (1.0–1.5 m) from the edge of the brooder to confine poults near the heat source. Guards can either be removed after a week or enlarged and retained for a second week.

Brooder guards should be 15–24 in. (40–60 cm) high. An open mesh material (mesh as large as three-fourths of an inch) such as lightweight hardware cloth or chicken wire can be used during warm weather. A solid material such as corrugated paper, tarred paper, or plywood should be used during colder weather. The solid wall protects poults from drafts and helps retain heat near the hover area. Brace the brooder guards as needed to keep them in place.

Many growers use double guards in which each ring encircles two brooder units to give poults more total area. Poults should be checked when double brooder guards are used to see that too many do not crowd under one hover.

EQUIPMENT ARRANGEMENT

Figure 13.5 illustrates a typical initial arrangement of brooding equipment. Pan-type automatic feeders are used in this system. Crinkled paper is rolled out under the brooders, and the first feed is placed on the paper. The paper can be rolled up and discarded 5–7 days after placement because the poults will be using automatic feed pans by this time.

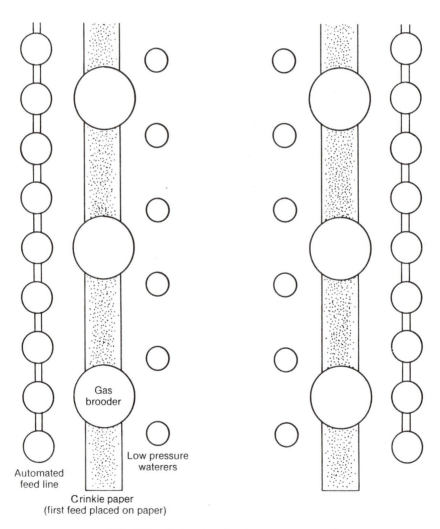

FIG. 13.5.   Turkey brooding house arrangement.

Water fountains should be filled and placed along the edge of the hover at least 12 hrs before the poults arrive so the water is at room temperature.

After 3 to 4 days waterers can be gradually moved away from the edge of the hover. The feed pans should be kept full for the first 4 days. The poults saved by providing them with easy access to feed will pay for any wasted feed. The level of feed in the pans can then be reduced to no more than half full to control feed wastage. After the first week, raise the feeders at weekly intervals so the lip of the feeder is the same height as the poult's back.

Adequate ventilation is critical in brooding poultry. One gallon of LP gas produces 1 gal of water and uses up considerable oxygen. Ventilation should be at least 0.2 ft^3/min per bird.

Table 13.2 lists the requirements and equipment recommended for brooding poults.

One problem peculiar to brooding turkey poults is the refusal of day-old poults to eat and drink. It is estimated that .5 to 3% mortality comes from this problem. Low temperatures, low light intensity, stale air, disturbing poults during their first 24 hr in the brooder houses, crowding cull poults, and poults too long in transit are all causes given for starveouts by producers.

Poults should always have feed and water within 10 to 15 ft. To prevent wastage, troughs or automatic feeder pans should be filled only half full. The Minnesota Agricultural Extension Service recommends feeding and water spaces as follows.

TABLE 13.2. Facility and Equipment Recommended for Brooding Turkeys

	Age of poult	
Item	0–4 weeks brooding	4–8 weeks brooding
Space	.5–.6 ft^2 (.05 m^2) per poult	1–1 ¼ ft^2 (.13 m^2) per poult
Brooders	1 per 250–300 poults	(Space heaters or brooders, as needed)
Waters		
Gallon jars	2 per 100 poults	—
Trough	1 in. per poult (2.5 cm per poult)	1 in. per poult (2.5 cm per poult)
Feeders[a]		
Trough	3 in. per poult (7.5 cm per poult)	3 in. per poult (7.5 cm)
Round	—	3 in. per poult (7.5 cm)

[a] Recommendations are based on linear inches of feeding space. Trough feeders provide feeding space on both sides, so a feeder 3 ft long (1m) provides 6 ft (2m) of feeding space. Measure the outer circumference of round feeder pans to determine available space. A 25- or 30-lb (12–14 kg) round feeder generally has a 16- or 17-in. (40–43 cm) diameter pan that is 50–53 in. (125–130 cm) in circumference.

TROUGH FEEDERS

First 2 weeks—160 linear ft (26 m) (80-ft troughs) of feeding space per 1000 poults.

3–4 weeks—240 linear ft (75 m) of feeding space per 1000 poults.

5–6 weeks—320 linear ft (112 m) of feeding space per 1000 poults.

7–8 weeks—480 linear ft (150 m) of feeding space per 1000 poults.

12 weeks on—560 linear ft (175 m) of feeding space per 1000 poults.

MECHANICAL FEEDERS

Hanging type—30 feeders per 1000 poults.

Trough type—80–100 linear ft (26–32 m) per 1000 poults.

A sudden drop in feed intake is a signal for a possible disease problem.

WATERERS

Minimum water capacity per 1000 poults: Additional water space may be needed during hot weather, but recommended levels are as follows:

	Fountains	Troughs
First 2 weeks	Twenty 1-gal	30 linear ft (9 m)
3–4 weeks		60 linear ft (18 m)
5–8 weeks		80 linear ft (26 m)

Automatic waterers: Use three 6-ft troughs per 1000 poults or one trough per stove after the first week.

Range tanks: Use one 200- to 300-gal tank (750–1125 liters) per 1000 poults.

At the end of the brooding period the young turkeys are either moved to fenced-in range or to growout houses for confinement rearing. As turkeys are grown more on a year-round basis, the trend is toward more and more confinement rearing.

REARING TURKEYS

In the southeastern United States about 72% of the turkeys are grown in confinement and 28% on open range (Fig. 13.6). Georgia

FIG. 13.6.   Some turkeys are still reared on open range.

growers report that under suitable weather conditions turkeys grow to heavier weights more efficiently on open ranges than in confinement houses.

About 1.5 acres of sloping, partially shaded, well-drained land per 1000 birds is required. Four manual range feeders and four tub waterers are needed for each 1000 birds. In a Georgia survey it was found that 46.5 hr of labor was required for toms (8–22 weeks) and 31.6 hr for hens (8–18 weeks) to grow turkeys on range. Labor requirements for confinement rearing were almost the same as for range rearing. Predators, diseases, and weather conditions, particularly early storms in the north, are some of the hazards of range rearing.

Confinement rearing is a recent trend in turkey production. Some of the advantages claimed are (1) better protection against thieves, predators, disease, and adverse weather conditions, (2) lower land costs, (3) lower labor cost because of automatic feeding and watering, and (4) better control of overall operations because the birds are confined closely together and are more dependent on the operator. Some disadvantages cited include (1) higher costs for housing and equipment, (2) greater risk of respiratory diseases and cannibalism, and (3) dangers from overcrowding.

In Georgia confinement houses are generally of pole construction, with sheet metal Styrofoam insulated roofs, side curtains, and dirt litter floors. Houses are generally 50 ft wide (16 m) and 400–500 ft

(130–160 m) in length. About 2.0–2.5 ft^2 (0.2–0.25 m^2) is used for hens and 3.0–3.5 ft^2 (0.3–0.35 m^2) for toms. Birds are fed by mechanical feeders. Wet spots in the litter are stirred as needed. Hens and toms are reared separately.

The costs of growing turkeys by independent growers in Georgia under various growout plans are compared in Table 13.3. Initial investment and annual cost for a 100,000 turkey operation are shown in Table 13.4. Table 13.5 shows data for housing costs in Georgia.

The turkey industry in recent years has closely followed trends in the broiler industry for fewer producers and larger flocks. For example, in a 1982 Georgia study of 79 producers, 50% of the producers grew less than 50,000 turkeys per year and 10% grew more than 500,000.

TABLE 13.3. Summary of Average Costs for Independent Turkey Growers by Types of Growout Systems Used in Georgia, 1981

	Total confinement		Open range	
Systems	Toms	Hens	Toms	Hens
Market weights[a] (lb)	26.80	14.00	28.50	15.40
Cost items (cents/lb)				
Poults[b]	4.53	5.59	4.37	4.87
Feed[c]	32.81	30.04	31.88	27.76
Building, equipment, depreciation, and interest[d]	1.21	1.74	0.58	1.05
Supplemental equipment and hauling turkeys[e]	1.52	1.96	1.51	1.91
Propane electric[f]	0.53	1.01	0.36	0.67
Litter	0.15	0.29	0.07	0.13
Medication	0.27	0.51	0.25	0.46
Insurance, taxes, and repairs	0.11	0.15	0.11	0.20
Truck and tractor fuel	0.15	0.29	0.14	0.26
Loading turkeys	0.15	0.29	0.14	0.26
Miscellaneous	0.04	0.07	0.04	0.07
Interest on operating capital[g]	1.48	1.45	1.43	1.38
Total cost excluding operator labor, land, and management[h]	42.95	43.39	40.88	39.02

[a] Market weights before condemnation losses were deducted. Market weights used for cost computations in the above summary after condemnation losses were confinement toms, 26.3 lb; confinement hens, 13.78 lb; range toms, 27,89 lb; and range hens, 15.2 lb.
[b] Tom poults valued at $1.04 and hen poults valued at $0.68.
[c] Feed valued at $200/ton.
[d] Source: Table 1.
[e] Custom hauling rate of $.01/lb (240-mile roundtrip) to the processing plant is added to the supplemental equipment cost from Table 2.
[f] Propane average 171 gal/1000 poults by full-house brooding. Electricity average 189 kWh/1000 poults for both types of systems.
[g] Interest on operating capital computed for 3 months/flock at 15% annual rate.
[h] Labor rates averaged confinement toms, 68 hr/1000; confinement hens, 55 hr/1000; range toms, 69 hr/1000; and range hens, 54.1 hr/1000.

TABLE 13.4. Initial Investment and Annual Costs for Supplemental Equipment 100,000-Turkey Operation, Georgia, 1981

Item	Number	Independent growers		Contract growers	
		Initial cost ($)	Annual repairs ($)	Initial cost ($)	Annual repairs ($)
Well and pump	1	2,518	107	2,518	107
Farm tractor[a]	1	1,358	246	1,358	246
Front end loader	1	1,314	98	1,314	98
Spreader truck	1	3,500	350	3,500	350
Feed truck	1	28,000	800	—	—
Sprayer	1	693	51	693	51
Pickup truck[b]	1	2,280	307	2,280	307
Mower	1	813	88	813	88
Farm trailer	1	2,080	10	2,080	10
Turkey loader	1	12,750	50	—	—
Scraper blade	1	208	—	208	—
Loading fans	4	800	—	—	—
Feeder auger wagon	1	1,100	60	1,100	60
Total		57,414	2,167	15,864	1,317

Annual fixed cost	100,000 turkeys	1,000 turkeys	100,000 turkeys	1,000 turkeys
Depreciation—10%[c]	5,741	54.41	1,586	15.86
Interest—15%[d]	4,306	43.06	1,190	11.90
Repairs	2,167	21.67	1,317	13.17
Insurance[e]	861	8.61	238	2.38
Taxes[f]	344	3.44	95	.95
Total annual cost	13,419	134.19	4,426	44.26

[a] Average price reported for used tractor was $5430 (25% charged to turkey operation).
[b] Average price reported for used half-ton truck was $4560 (50% charged to turkey operation).
[c] Depreciation computed over 10 years by straight-line method.
[d] Interest computed on average amount of loan over 10 years at 15% per year.
[e] Insurance cost estimated to be 1.5% of average purchase price.
[f] Taxes estimated for 40% of average value over 10 years at 30 mills.

Each year there are fewer independent producers because the production of turkeys, like other types of animal agriculture, requires the capital expenditure of large sums of money, assumption of high risk, uncertainty in the market place, and only a modest return on investment. The majority of turkeys are raised under contract or some similar marketing arrangement.

Producers who rear turkeys under contract are paid according to the performance of the flock compared to other flocks grown at that time. Feed conversion and average weight of the flock are keys to satisfactory returns to the producers. Other arrangements may compare total cost per pound of processed turkey produced.

TABLE 13.5. Turkey House, Equipment, and Range Construction Costs, Georgia, 1981 [a]

Type facility	Number needed	House size (ft)	Area (ft)	Capacity (number of birds)	Building and equipment costs ($)			
					Construction	Equipment	Total	ft²
Brooder house [b]	1	40×300	12,000	12,000 [e]	18,720	10,069	28,789	2.40
Growout house [c]	2	50×400	40,000	12,000 toms/ 16,000 hens	67,200	13.220	80,400	2.01
Range [d]	1	—	15 acres	10,000 [e]	1,940	4,000	5,940	—

[a] Source: Lance (1983).
[b] Conventional open-sided house with insulated roof. Equipment included gas brooders, automatic poult feeders, drinkers, minidrinkers, feed bin, curtain machine, and curtains. Brooder house for 16,000 hen poults (40 × 400 ft) cost $38,400.
[c] Conventional open-sided houses with insulated roofs. Equipment included automatic adult turkey feeders, drinkers, feed bin, and curtains.
[d] Range investment included stock wire, posts, water pipe, 1000-lb feeders, and drinkers. Feeders and drinkers are moved among ranges as new flocks are started. Range land in Georgia was renting in 1982 for approximately $25/acre.
[e] The same capacity was usually used for toms and hens.

The turkey industry has expanded rapidly in the last 15 years. Consumption of turkey has increased from 7.9 to 10.8 lbs per capita. The expansion in consumption has come from further processed turkey items not the sale of whole frozen turkeys. Further processing has several advantages for the turkey industry. An obvious advantage is a more uniform flow of product in marketing channels throughout the year. Processors add value to the turkey product by additional processing such as cooking so the product will sell at a higher price and a larger profit.

Turkeys were once reared in many states in relatively large numbers. In 1986, North Carolina, Minnesota, California, Arkansas, Missouri, and Virginia were the leading turkey-producing states (Table 13.6).

TABLE 13.6. Rank of States in Turkey Production, 1966–1986

State	Year	Rank	Turkey production (in millions)
North Carolina	1966	9	5.3
	1976	3	16.7
	1986	1	39.1
Minnesota	1966	1	16.4
	1976	1	24.4
	1986	2	34.2
California	1966	2	16.9
	1976	2	17.5
	1986	3	21.9
Arkansas	1966	5	6.5
	1976	4	10.1
	1986	4	16.5
Virginia	1966	7	6.2
	1976	7	7.3
	1986	5	13.8
Missouri	1966	3	10.0
	1976	5	9.7
	1986	6	13.5
Indiana	1966	10	4.2
	1976	9	5.2
	1986	7	9.3
Pennsylvania	1966	15	1.9
	1976	11	3.7
	1986	8	7.8
Iowa	1966	4	7.2
	1976	8	6.5
	1986	9	7.0
Wisconsin	1966	8	5.5
	1976	10	5.1
	1986	10	6.1
Texas	1966	6	6.3
	1976	10	82.2
	1986	11	N/A

Source: For 1966 data: USDA (1967) Statistical Reporting Service, POU 2-3(67), 10.
For 1976 data: USDA (1977) Statistical Reporting Service, Statistical Bulletin No. 677.
For 1986 data: USDA (1987) Statistical Reporting Service, POU 3-1(87).

The future appears bright for the turkey industry. Improvements in housing, nutrition, genetics, and disease control should continue to improve the efficiencies of producing turkeys. The acceptance by the consuming public of many of the further processed products made from turkey meat also means expansion of the production side of the industry.

Dressing yields in turkeys surpass chickens by a considerable margin. Eventually, mechanical deboning of whole turkeys will be perfected. Because turkey meat, the same as other poultry meats, is relatively bland, it can be seasoned and formulated into numerous products. Turkey meat is also delicious and lends itself well to a variety of menus. Turkey appeals to a diet-conscious public because of its high nutritional value, low fat content, and affordable price.

REFERENCES

Anonymous. 1978. Turkey Breeding Manual. Nicholas Turkey Farm, Sonoma, CA.

Arrington, L. C. 1980. Market Turkey Management—Brooding. North Central Regional Ext. Publ. 115, Univ. of Wisconsin, Madison, WI.

Berg, R. W. 1976. Early Development of Turkey Breeders, Gobbles. Minn. Turkey Growers Assoc., St. Paul, MN.

Brewer, C. E., and Mills, W. E. 1979. Brooding Turkeys. PS&T Guide No. 8, North Carolina State Univ., Raleigh, NC.

Lance, G. C. 1982. Production and Marketing Contracts for Georgia Turkey Producers. Univ. of Georgia Res. Bull. 280, Athens, GA.

Lance, G. C. 1983. Production Costs and Returns for Independent and Contract Turkey Growers in Georgia. Univ. of Georgia Res. Bull. 301, Athens, GA.

Lewis, J. C. 1973. The World of the Wild Turkey. J. B. Lippincott, Philadelphia, PA.

Marsden, S. J. 1971. Turkey Production. USDA, Agricultural Research Service, Washington, D.C.

Olsen, S. J. 1968. Fish, Amphibians and Reptile Remains from Archaeological Sites. I. Peabody Museum, Harvard University, Cambridge, MA.

Skinner, J. L., and Arrington, L. C. 1969. Artificial Insemination of Poultry. Fact Sheet 31, Univ. of Wisconsin, Madison, WI.

USDA 1967. Statistical Reporting Service, POU 2–3 (67).

USDA 1977. Statistical Reporting Service, Statistical Bulletin No. 677.

USDA 1987. Statistical Reporting Service, POU 3–1 (87).

14

Waterfowl Production

Ducks, geese, and swan are part of a large classification of waterfowl, known as Anatidae, easily identified by their heavy bodies, short legs, and webbed feet. Their bills have a hard horny mail at the tip with transverse tooth-like ridges on the biting edge. Waterfowl as the name implies reside in and around water. They have precocial young, are down covered, can run about at hatching, and feed themselves.

DUCKS

The word duck has become a term used to describe the Anatinae, a branch of Anatidae. Only a few members of the Anatidae have been domesticated and commercially produced for meat, eggs, and feathers. The Anatidae family can be distinguished from other waterfowl by their smaller size, shorter necks and legs, flatter bodies, and broader bills (Fig. 14.1).

Male ducks are known as drakes while the females are known as ducks or duck hens. The young, newly hatched and immature ducks are referred to as ducklings. Most of today's domestic breeds of ducks appear to be descendents of the wild mallard *(Anas platyrhynchos)*.

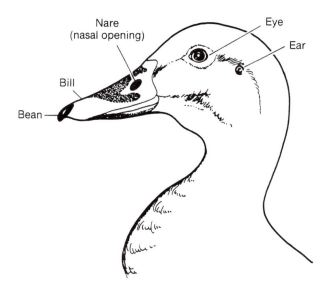

FIG. 14.1. Anatomical features of the duck.

All domesticated true ducks retain the distinctive crescent sex feathers seen at the upper portion of the tail of the drakes. The voice of the drake is a hoarse husky whisper while that of the hen is a loud, rather flat quack. With experience, ducks can be sexed by their voice and by the appearance of the "drake" feathers of the tail. Drakes also develop larger and coarser heads than duck hens.

The Duck Industry

The commercial duck industry in the United States is focused around the production of young ducklings for meat purposes. The duck industry, the same as the broiler and turkey industry, prefers a white-feathered duck. Dark-colored down missed in processing results in a less desirable carcass appearance. The United States duck industry utilizes the White Pekin type duck (Fig. 14.2).

In the United States duck production in 1984 was only 12 million birds. The industry has grown slowly for several reasons. American consumers tend to be oriented toward chicken and turkey with the majority of the public choosing white meat over dark meat. When consumers try duck meat they are in many cases skeptical of the darker meat. Also, because of the fat covering on ducks the yield of edible meat is less than for turkeys and chickens. Lack of acceptance has also been shown to relate to the fact that the general public is unfamiliar with the proper methods of preparing duckling. In addition to these factors, the meat of ducklings is more expensive than that of

FIG. 14.2. The United States commercial duck industry utilizes the White Pekin breed of ducks that originated in China.

chickens and turkeys. Industry leaders believe that to be successful the duck industry will have to carefully advertise its product and help educate the consumer in proper preparation methods.

Duckling is on the menu in many of the better restaurants throughout the country. Consumption has also benefited from the increase in popularity of oriental restaurants.

In retail markets ducks are generally sold frozen and tend to sell better in large cities in which there are ethnic groups familiar with utilizing duckling in the preparation of various dishes. Fresh marketing has been limited by a lack of sale volume.

Duck production was once centered on Long Island, New York, the origin of the restaurant trade term "Long Island Duckling." Long Island still produces ducklings but human population pressures have raised costs and caused production declines in the area. Ducklings are now reared and processed in large numbers in Indiana, Michigan, North Carolina, and Minnesota, as well as New York.

Meat-Type Ducks

The United States commercial duck industry utilizes the White Pekin breed of ducks that originated in China. This breed was brought to the United States in the 1870s to Long Island, New York. The Pekin is well suited for commercial purposes since it is white plumaged, matures rapidly, and attains a mature weight of 9 lb (4.1 kg) for drakes and 7.5–8 lb (3.4–3.6 kg) for hens. Pekins are also good egg producers and do not become broody easily while producing 150–180 eggs per year. Pekin ducklings grow rapidly and can be processed at 7 weeks of age.

Another duck raised in small flocks that has good meat qualities is the Muscovy *(Cavina moschata)*. The Muscovy is not a true duck in the general sense of the designation. It originated in South America and is a tree nesting (arboreal) duck that can easily be identified by the fact that it has caruncles on its face and upper neck much like a turkey. It does not have the characteristic crescent sex feathers found in the breeds and varieties, such as the Pekin that developed from the wild mallard *(Anas platyrhynchos)*. Recognized varieties of Muscovy ducks can be black, white, blue, or chocolate. Lighter colored varieties are easier to process and all muscovies have white skin. They weigh 7–10 lb (3.2–4.5 kg) as adults and are not as prolific egg layers as Pekins. Muscovy ducks are more active than Pekins and do not carry as much fat. The fact that their carcass is leaner makes them more attractive to some consumers.

Muscovy ducks have not experienced commercial acceptance as a meat bird in the United States. Muscovy ducks grow slowly compared to Pekins and are not ready to process until they are 12–14 weeks of age. This makes them uneconomical to produce unless a specialty market for their leaner carcass exists. In commercial rearing situations Muscovy ducks are more quarrelsome and aggressive than other ducks. A problem experienced in commercial production has been cannibalism when they are overheated or overcrowded.

Muscovy ducks are hardy, quite disease resistant, and good foragers when kept loose in small flock situations.

Muscovy ducks have the same chromosome number (80) as other ducks but when crossbred with other ducks produce sterile hybrids or "mule" ducks. Generally, "mule" ducks are produced by crossing muscovy hens with mallard-type drakes. The crosses are being produced on a limited basis for the restaurant trade. The breasts of "mule" ducks are commonly filleted, marinated, then rolled and cooked like a fillet mignon.

Other meat-type ducks include the Aylesbury, a large deep-bodied breed developed in England and popular in Europe. It is similar in appearance to the Pekin with several exceptions: the Aylesbury is deeper bodied, has white skin, and a flesh-colored bill. The legs and feet are lighter in color than Pekins. Commercial duck breeders often breed Aylesburys with their Pekin stock.

Other ducks large enough for meat production include the Rouen, Cayuga, and Swedish breeds. In the United States, the colored plumage and poor egg-laying ability of these breeds have relegated them to the interest of hobbyists exhibiting ducks.

Egg-Type Ducks

The most popular breed of egg-type duck is the Khaki Campbell (Fig. 14.3). The Khaki Campbell has been described as the leghorn of waterfowl in recognition of its egg-laying ability. The Khaki Campbell originated in England and reportedly is the result of a cross involving the Fawn and White Runner, Rouen, and Mallard ducks. Drakes are brownish-bronze over the tail coverts, heads, and necks with the remainder of the body khaki brown. The drakes have greenish bills

FIG. 14.3. The most popular egg-type duck is the Khaki Campbell.

and orange feet and legs. Khaki hens have brownish heads and necks with the rest of the plumage khaki brown. The hens have greenish dark bills and brown feet and legs.

Khaki Campbells are not suited for meat production because young drakes and hens only weigh around 3 lb (1.4 kg) at 7 weeks of age. Adult Khaki Campbells weigh in the range of 4–4.5 lb (1.8–2 kg). Selected strains of Khaki Campbell ducks have laid close to 365 eggs per hen within a laying year, outperforming the best strains of chickens.

The Indian Runner originated in the East Indies but the egg-laying ability of the breed was later developed in Europe. Indian Runners are surpassed only by Khaki Campbell ducks in egg-laying ability. Under good management conditions, egg numbers per duck in excess of 250 per year are not uncommon. The New Standard of Perfection recognizes eight varieties of Indian Runners. The Fawn and White, the Penciled, and the White Runner are the most common. Runner ducks of good quality stand almost erect with a nearly perpendicular carriage. Runner ducks are small for meat purposes with mature weights of only 3.5–4.5 lb (1.6–2.0 kg).

Runner ducks are able to maintain egg production better than other ducks in situations in which access to water is limited.

In spite of the good egg-laying ability of both the Khaki Campbell and Indian Runner ducks, several problems exist. Generally, in confinement production systems, the feed conversion per dozen eggs of duck hens is considerably higher than that of chickens. When compared to chickens, ducks are messy and are harder to manage when kept on litter. When attempts have been made to cage duck hens like chicken hens, the ducks developed serious foot and leg problems because of their webbed feet.

Duck eggs are larger in size than chicken eggs, have thicker and stronger shells, have tougher shell membranes, and possess albumen that is more firm. They can be used for cooking with excellent results. However, most American consumers have not accepted them for use as table eggs.

Egg-type ducks appear to have considerable potential in developing countries or in small flock situations. Rearing ducks for egg production in Europe is popular. This is also true for the Far East where duck eggs are used in making balut, a fermented specialty utilizing embryonated eggs.

Bantam Ducks

Several breeds of ducks have been discussed that have meat or egg production potential. In the United States, Bantam ducks are becoming increasingly popular to raise for exhibition and other aesthetically pleasing experiences for the hobbyist. Popular breeds of these ducks include English Gray Calls, White Calls, and Black East Indies.

Breeder Stock

When duck breeding stock is selected, individual birds are chosen on the basis of their weight, conformation, and complete feathering. These candidates should come from parent stock demonstrating high fertility, good hatchability, and a high rate of lay. Ducks are monogamous when they are wild but under domestic conditions they become polygamous. Domestic ducks are normally mated six hens per drake. Ideally, drakes should be slightly older than the hens. Ducks mature slightly slower than chickens. Ducks should not be brought into production before 7 months of age because immature birds lay smaller eggs and the fertility of these eggs is low. Duck hens are normally stimulated 3 weeks before egg production is desired; it is desirable to light drakes 1 or 2 weeks before the hens to ensure good fertility. Ducks can be light stimulated with 14 hr of light per day. After egg production starts peak production should occur in 6 weeks and reach 90% or better production. Meat-type ducks can sustain production above 50% for 5–6 months. Their fertility generally peaks close to the time that egg production peaks. Ducks can be induced to molt and can be recycled like chicken and turkey hens if it appears economically feasible for the producer. When recycling is practiced, it is best to replace the males with young drakes.

Breeder Facilities

Although most producers provide swimming water for breeding stock, it is not essential for the production of fertile eggs. Ducks generally mate in water but will mate in other places. Breeder ducks are usually confined to the laying houses only at night. During the day they are turned into runs off the buildings or into waterways. Breeder ducks normally have 4–5 ft^2 space per duck (0.37–0.46 m^2). Partly littered floors have proven more satisfactory than complete slats or wire floors. To maintain sanitation it is necessary for houses to be kept well ventilated and as dry as possible. Most producers find it necessary to add litter from time to time and to remove damp litter periodically.

Migratory waterfowl such as ducks evolved in a feast-or-famine environment. When feed is supplied freely, ducks tend to overeat and become obese. Obesity can cause reproductive problems. For this reason, breeder ducks are placed on restricted feeding programs during the growing period and in the period after they peak in production. Breeders require a quality ration of 16–18% protein. To avoid waste, the ration must be pelleted. The goal in feeding breeder ducks is to achieve maximum fertility and egg production while maintaining body weight without appreciable addition of body fat. A rough estimate for feed consumption of breeder ducks is 44 lb (20 kg) per 100 breeders divided into a morning and an afternoon feeding. It is important that the feed is distributed so that all of the breeders have access to the feed at the same time, otherwise aggressive birds tend to overconsume.

Most ducks lay their eggs early in the morning. This allows producers to confine breeders at night and then turn them into yards or runs during the day. Ducks will make nests in the litter but will utilize nest boxes placed at floor level. These nests are usually made in long rows and placed away from the walls. Nests are generally constructed 12 in. wide, 18 in. deep, and 12 in. high (30 × 45 × 30 cm). It is common practice to provide boxes open on two sides. After the eggs are laid ducks commonly attempt to cover them with litter. For sanitary purposes the litter in nest areas should be kept clean and fresh.

Egg Handling

Producers find it best to gather eggs early in the morning. This limits the amount of soiling and reduces the number of cracks. Duck eggs are washed in water of 110–115°F (44–46°C), then rinsed in water that is slightly warmer. Abnormal, misshapen, or cracked eggs are discarded. Hatching eggs are stored small end down at a temperature of 55°F (17°C) and a relative humidity of 75%. If weekly egg settings are made the eggs do not need turning during storage. Hatching eggs can be stored up to 2 weeks but after the first week, the eggs should be turned daily to reduce declines in hatchability.

Incubation

The eggs of Muscovy ducks require 35 days for incubation; other domestic ducks require approximately 28 days depending on the size of the egg. Incubators are available that are especially designed for waterfowl eggs. Incubators designed for chicken and turkey eggs can be used if the directions of the manufacturer for their use with duck eggs are carefully followed. The major difference will be in the level of humidity, not in temperature. The eggs of waterfowl benefit when the angle or degree of turning is increased over that used for the eggs of turkeys and chickens. Duck eggs are routinely fumigated within 12 hr after setting to prevent bacterial contamination of the egg contents and the passage of bacteria through the pores of the shell. Duck eggs incubated in forced air incubators at 99–100°F (37.4°C) are held at a wet bulb reading of 86°F (30°C) during incubation; on the twenty-fifth day the eggs are transferred to a hatching compartment (Fig. 14.4). At this time the humidity is raised gradually to a 93° wet bulb reading by

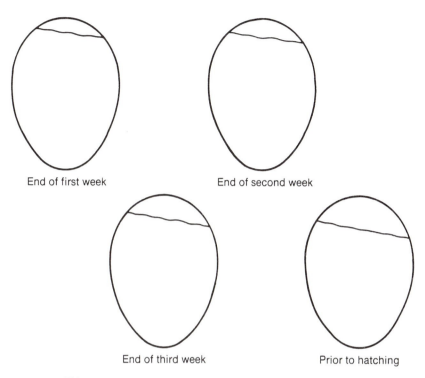

End of first week   End of second week

End of third week   Prior to hatching

FIG. 14.4.   Air cell changes during incubation of waterfowl eggs.

the end of the twenty-sixth day. At the end of the twenty-seventh day, when the ducklings have just about completed their hatching, the humidity is reduced to 90°.

Brooding

The brooding period is the early part of the life of the duckling when supplemental heat is supplied to aid in maintaining body temperature until the duckling grows feathers. Any type of poultry building can be used as long as the ducklings are kept warm, dry, and away from drafts.

Adequate ventilation is needed. The system must be designed not to chill the ducklings. Floors can be litter or wire in duck growout houses. Wire floors are made of welded wire (0.75 in.) (1.5 cm) placed on a frame about 4 in. (10 cm) above concrete. This house type can be washed down and drained into a lagoon or similar waste disposal system. The advantages of wire flooring are that ducklings are kept away from manure and dampness; however, the cost of this type of house is becoming prohibitive. Litter floors require good management and litter material must be free of the mold spores of *Aspergillosis flavus*.

Ducklings require adequate floor space. Densities of 0.5 ft^2 (.05 m^2) on wire and 1.0 ft^2 (.1 m^2) on litter are generally used to 3 weeks of age. If total confinement is practiced to 7 weeks of age, floor space of 2.0–2.5 ft^2 (0.2–0.25 m^2) per duck is required.

Generally, ducklings need a source of brooding heat for about 4 weeks. Initially, a brooder temperature of 85–90°F (30–32°C) is adequate. The temperature is dropped 5°F (3°C) per week. Ducklings consume large quantities of water; in commercial housing it is supplied by automatic waterers. Ideally, the waterers are placed over wire flooring or near a screened drain. The waterers must be cleaned regularly to help prevent disease.

Nutrition

Ducklings are fed a high-energy, high-protein starter feed for the first 2 weeks. The feed is fed as crumbles or small pellets. Feed in this form is easier to consume, reduces waste, and results in improved feed conversion. A grower feed of ⅛-in. pellets is used until marketing. Examples of duck rations are given in Table 14.1.

Health Management

Ducks are fairly resistant to disease and muscovies are more resistant than Pekin or Runner types. Ducklings reared for market purposes are not vaccinated unless problems have been experienced on the farm or in nearby areas. Diseases that can affect ducks include botulism, fowl cholera, necrotic enteritis, virus hepatitis, aspergillosis, infectious serositis, and duck virus enteritis. Many of these diseases are discussed in Chapter 9. Commercial duck producers avoid many of these diseases by utilizing good management, proper sanitation, and confinement and isolation of their flocks.

GEESE

Goose, the plural of which is geese, is a term used to describe members of a subfamily (Anserinae) of a classification of waterfowl called the Anatidae. Geese are intermediate in type and size between ducks and swans. They have relatively long necks, their face areas are feathered, and their shanks are reticulated. In the wild, the sexes are identical in plumage color, making sex determination difficult without handling. (Figure 14.5).

In geese, the male is called a gander, the female is known as a goose, and young geese are called goslings. Geese are believed to be the first

TABLE 14.1. Examples of Typical Duck Rations (kg) [a]

Ingredient	Starter	Grower	Holding	Breeder
Ground yellow corn	412	—	445	518
Ground wheat	—	58	—	—
Ground barley	100	100	114	100
Wheat shorts	50	50	200	50
Wheat middlings	50	50	—	50
Dehydrated greenfeed	20	—	120	20
Meat meal (50%)	—	—	20	—
Fish meal (60%)	—	—	—	—
Soybean meal (49%)	332	178	60	195
Ground limestone	11	12.5	17.5	44.5
Calcium phosphate (20%)	12.5	13	11	10
Salt (iodized)	2.5	2.5	2.5	2.5
Vitamin-mineral premix	10	10	10	10
DL-methionine	0.50	0.25	0.50	—
Choline chloride	—	—	—	0.13
Total	1000	1000	1000	1000
Crude protein (%)	23.0	19.0	13.7	17.0
Crude fat (%)	2.5	1.6	3.8	2.8
Crude fiber (%)	3.8	3.5	5.9	3.7
Metabolizable energy, kcal/kg	2761	2805	2541	2756
(kcal/lb)	(1253)	(1275)	(1155)	(1253)
Calcium (%)	0.79	0.82	1.10	2.00
Available phosphorus (%)	0.39	0.39	0.40	0.40
Choline equivalents, mg/kg	1731	2264	2070	1571
(lb)	(787)	(1029)	(941)	(714)
Methionine (% of diet)	0.40	0.30	0.25	0.28
Cystine (% of diet)	0.33	0.29	0.18	0.25
Lysine (% of diet)	1.20	0.95	0.54	0.85
Tryptophan (% of diet)	0.30	0.28	0.20	0.23
Threonine (% of diet)	0.89	0.67	0.48	0.69

[a] Source: Summers and Leeson (1976).

domestic poultry kept by man. They are intelligent birds, learn quickly, and demonstrate good memory.

Domestic ganders become aggressive during the breeding season. This aggression increases as they age. Ganders can inflict injury by biting and giving heavy blows with their wings. Geese are a gregarious bird and congregate in flocks except during the spring breeding season. They generally have extended longevity compared to other poultry, with life spans of 20 to 60 years not uncommon; however, they are seldom kept as breeders for more than 4 or 5 years. Geese differ from other poultry in respect to their characteristic of grazing. Geese are monogastrics and cannot digest cellulose, so they graze on tender vegetation, preferring grasses or legumes such as alfalfa.

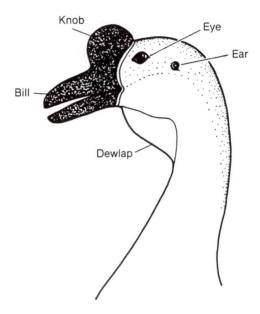

FIG. 14.5. Anatomical features of the goose.

The Goose Industry

Geese were popular on farms in the United States at the turn of the century with more geese raised than chickens or turkeys (Fig. 14.6). Today less than a million geese are raised commercially in the United States per year. The leading states are South Dakota, Iowa, and Minnesota. However, geese are more popular in Canada and considered the holiday bird of choice in much of Europe. Leading goose-producing countries include Poland, Hungary, Czechoslovakia, France, and Bulgaria. Many of the geese in Europe are grown in relatively small flocks on grasses and then fattened for slaughter.

Goose production has been hampered in the United States by some of the same problems that face the duck industry. Namely, United States consumers are unfamiliar with preparing geese, prefer white meat, and are unwilling to pay higher prices for geese when chicken and turkey are available in abundant supplies at lower prices. A National Goose Council has been formed to help promote the goose industry.

Over the past 50 years major progress has been made in improving the performance of chickens, turkeys, and ducks by manipulating breeding, selection, nutrition, management, and disease control. Geese have practically been neglected during this time and yet the goose is still the fastest growing bird to 10 weeks of age, often exceeding 11 lb (5 kg) by that time.

FIG. 14.6.   A flock of young geese not ready for market.

Heavy Geese

The commercial goose industry in the United States is based almost entirely on the production of the heavy and medium classes of geese listed in the American Standard of Perfection. All of these breeds of geese trace their ancestry to the Gray Lag goose of Europe *(Anser anser)*.

Several breeds of heavy geese are grown for processing purposes. Processors like the processors of other poultry prefer light-colored birds because of the problem associated with dark pinfeathers.

The Embden or White Embden goose is a large deep-bodied goose that traces its history to goose stocks in Germany and England. The modern day Embden was developed in England and deserves listing as the top of all domestic breeds of geese. Embdens weigh 25 lb (11.3 kg) or more and goose hens weigh 20 lb (9.1 kg). Embdens lay 20–40 eggs per season depending on the age of the breeders.

Gray is the most common color of the domestic breeds of geese descending from the European Gray Lag goose. Many novice goose breeders classify geese by color. Many gray-plumaged geese are called Toulouse geese even though they only faintly resemble the Standard Toulouse. Many commercial geese are really crossbred geese with some Toulouse in their ancestry. The Toulouse was developed in England in response to a need for a large, easily fattened goose that had a large liver. Toulouse geese are better layers than Embden and some hens may lay as many as 60 eggs per season. This is why many breeders use Toulouse-type hens in their breeder flocks. Toulouse geese weigh the same as Embdens, 25 lb (11 kg) for adult ganders and 20 lb (9 kg) for hens.

The African goose follows the Embden and Toulouse in size. However, the African goose is related to the *Anser cygnoides* (wild China swan goose) rather than *Anser anser*. Little is known of its origin except that it has never been closely associated with the African continent.

It appears that exotic names were used to call attention to new breeds of geese and this appears to be the case in this situation. African geese can be recognized by their size, prominent dewlap, knob on the base of the bill, and their black bill. Adult ganders weigh in excess of 24 lb (11 kg) and hens 18 lb (8 kg). African geese are poorer layers than Embden or Toulouse and less tolerant to cold. African geese in the United States are kept primarily as exhibition waterfowl.

Medium Geese

Medium class geese yield attractive dressed carcasses and compete well in the junior goose market. American Buff geese can actually claim United States origin. They were developed from Pomeranian stock. American buffs now breed true and have a uniform buff coloring and orange to reddish-yellow feet and shanks. Adult ganders weigh 18 lb (8 kg) and hens 16 lb (7 kg). Buff geese lay 20–40 eggs per season depending on their age. Another relatively new breed of domestic geese is the Pilgrim. This breed was originated in Iowa by Oscar Grow. It is unique because since the breed exhibits sex linkage pure stocks are auto-sexing. Ganders are white and goose hens are gray. Ganders can be identified from white breeds because they are gray on the saddle of the back. Adult ganders weigh 16 lb (7 kg) and hens 13 lb (6 kg). Pilgrim geese are good layers and the goslings are hardy. Other medium geese include the Sebastopol, an exhibition and aesthetically pleasing novelty breed, and the Pomeranian. Pomeranian geese are of German origin and were developed as a commercial breed. In the United States they have remained an exhibition breed.

Light Geese

The most popular light goose is the White Chinese. Like the African goose, Chinese geese descended from the white China goose *(Anser cygnoides)*. Chinese geese fit both the ornamental as well as commercial category. China geese are preferred by commercial hatcherymen because they are prolific egg layers and the eggs hatch well when incubated artificially. China geese are probably one of the most practical breeds of waterfowl. China geese can be recognized by the slender neck and the upright stance and knob at the base of the upper mandible. The Standard of Perfection recognizes both a White China (Chinese) and Brown China goose. The white variety is easily recognized with the white plumage, knob, and orange bill and feet. The Brown China has a dark knob, black bill, and russet-brown plumage on the back and head. Adult ganders of both varieties weigh about 10–12 lb (4.5–5.5 kg). Good strains of China geese lay 40–75 eggs per season depending on their age and the management they receive.

Another less popular light breed of geese is the Tufted Roman goose, which is used primarily for exhibition.

The Canada goose is gaining popularity as an ornamental goose (Fig. 14.7). The Canada goose *(Branta canadensis)* is not really a goose but is a member of the brant family. Brants can be distinguished from true geese by the fact that they have black bills, feet, and shanks. Canada geese are handsome birds that mate in pairs. They have little utility and are poor egg layers, laying only 6–8 eggs per year. Under good conditions they reach sexual maturity between 2 and 3 years of age. The most common captive subspecies is the Eastern Canada goose. United States propagators are required to secure a propagation or a holding permit from the United States Fish and Wildlife Services because they are native migratory waterfowl. The Eastern gander weighs around 12 lb (5.5 kg) and the goose hen around 10 lb (4.5 kg).

The Egyptian goose *(Alopochen aegypticus)* is not a true goose but is a Shelduck. It is the smallest of the so-called geese and has a rather

FIG. 14.7. The Canada goose is rapidly gaining popularity as an ornamental goose.

exotic appearance with some irridescence in the plumage. It is also a poor egg layer, laying only 6–8 eggs per year. Egyptian geese mature between 2 and 3 years of age. As the name implies, they are native to the Nile region of Egypt. In America outside of the deep southern states, Egyptian geese need some protection or minimal housing during winter.

Breeding

Domestic geese differ from the other poultry previously discussed in their breeding habits. Geese can be successfully kept as breeders for 3 or more years. Goose hens generally peak in egg production during the third year and then start to decline in production. Geese in the wild mate in pairs but domestic geese can be mated with up to five females or hens per gander. Since only minimal attempts have been made in the United States to develop primary breeding stocks, commercial hatcherymen select breeders from flocks they have grown or have on contract.

Domestic ganders are harem oriented and the best fertility is usually acquired when harems can be formed 2 to 3 months before the breeding season. Adding or moving breeding stock at the onset or during the breeding season generally results in decreased fertility. Depending on the area of the country, geese on natural light will start laying in February in the south and not until April in the north. Breeder geese require minimal housing during the winter as long as adequate food and water can be provided. Breeder geese can forage for part of their nutritional needs on rye during the fall and early winter, fescue or orchard grass in the spring, and sudan grass in the summer to maintain their condition. Breeder geese are supplied with an 18–20% protein ration prior to and during the breeding season. Seventy-five to 125 lb (34–57 kg) of feed daily per 100 breeders is provided depending on the type of the breeder geese and their access to a grazing area. A method of limiting feed is to restrict the feed to the amount the breeders can consume in 10–15 min twice a day. Overweight breeder geese are not satisfactory because obese goose hens lay increased numbers of soft-shelled and double-yolked eggs. Fat geese also experience broken-down abdomens. To support egg production clean drinking water is provided at all times. Nest boxes are not used by many producers of hatching eggs. Instead, adequate amounts of straw or hay are placed around the edges of the pens so the hens can build their own nests. Most hens lay every other day because calcification of the shell takes longer than in other species of poultry. Some selected strains of China geese may lay more frequently. When housed breeder geese require at least 5 ft^2 (0.5 m^2) per bird. When breeder geese are in lots they are given 50–100 ft^2 (5–10 m^2) per bird or more depending on the vegetation. Grit and oyster shell supplied freely are beneficial to breeder geese.

Egg Handling and Incubation

Goose eggs are gathered at least twice a day to prevent breakage and soiling of the eggs as well as to prevent broodiness. Eggs are normally held at temperatures of 45–50°F (7 10°C). The egg storage room should have a relative humidity of 75%. Goose eggs can be held in storage up to 2 weeks, but shorter storage times tend to produce higher hatchability. Goose eggs normally require 28–30 days of incubation. The larger eggs of the larger breeds require a longer incubation than the smaller eggs of the small breeds. Forced air incubators are used in commercial goose hatcheries. Differences of opinion exist as to the proper method of incubating goose eggs. A temperature setting of 99–99.5°F (37.5°C) produces good results when humidity is kept at 86–88°F (30–31°C) wet bulb reading and the eggs are turned 180° eight times per day. Incubator manufacturers have incubators designed for goose eggs that can rotate the eggs 180°.

Goose egg incubation problems occur frequently in the south even when the instructions of the manufacturer of the incubator have been followed. When problems arise consideration should be given to the possibility that the humidity was so high that too little water was lost through the shell. A good rule of thumb for goose eggs is that at the time of hatch, the air cell should occupy 20% of the interior of the shell. In some instances the incubator operator may find it necessary to adjust the incubator humidity below manufacturer's recommendations to achieve successful hatches.

Since geese lay a limited number of eggs, goslings are expensive compared to most other poultry. Commercial goslings are priced at $250 to 350 per hundred depending on the type and quantity purchased.

Brooding

Goslings are best placed on feed within 24 hr after hatching. Care must be taken in the hatcher and when brooding to ensure that goslings can gain firm footing on nonslick materials. This prevents the goslings from becoming spraddle legged. Goslings need 1 ft^2 (0.1 m^2) of brooder space during the 3- to 4-week brooding period. A brooding temperature of 88–90°F (31–32°C) is used initially and reduced to 80°F (27°C) by the middle of the second week, 75°F (24°C) during the third week, and then if the brooding space is reasonably warm, no supplemental heat is required. Goslings can be started on any crumbled nonmedicated poultry starter ration of 18–20% protein. Higher protein rations generally cause more leg problems because they cause more rapid growth. Goslings tend to have higher requirements for niacin and riboflavin than other poultry. This requirement can be met by using Brewer's yeast when formulating rations for starting geese.

After brooding, goslings are normally allowed gradual access to pastures. This practice helps lower feed costs. Depending on the pasture vegetation, soil fertility, rainfall, and other variables up to 50 young geese or 20–25 adult geese can be grazed per acre.

Goslings are ready for processing at about 14 weeks of age. At this time they have completed a molt and can be processed without numerous pinfeathers. Geese that have been on pasture grazing since brooding are often processed at around 22 weeks of age. In most cases these geese must be put on a fattening ration 4 weeks prior to processing (Table 14.2).

TABLE 14.2. Examples of Typical Goose Rations (kg)[a]

Ingredient	Starter	Grower	Holding	Breeder
Ground yellow corn	432	613	390	552
Ground wheat	—	100	—	—
Ground barley	100	50	100	100
Wheat shorts	50	50	200	50
Wheat middlings	50	—	—	50
Dehydrated greenfeed	20	—	200	20
Meat meal (50%)	—	—	20	—
Fish meal (60%)	—	—	—	—
Soybean meal (49%)	310	150	45	158
Ground limestone	11	9.5	15	44
Calcium phosphate (20%)	13	12.5	11.5	11
Salt (iodized)	5	5	8.5	5
Vitamin-mineral premix	10	10	10	10
DL-methionine	—	0.38	0.50	0.25
Choline chloride	—	—	—	0.13
Total	1000	1000	1000	1000
Crude protein (%)	22.0	18.0	13.4	15.5
Crude fat (%)	2.6	1.7	3.8	2.9
Crude fiber (%)	3.8	3.5	5.9	3.7
Metabolizable energy, kcal/kg	2766	2818	2420	2776
(kcal/lb)	(1257)	(1281)	(1100)	(1262)
Calcium (%)	0.79	0.71	1.10	2.00
Available phosphorus (%)	0.39	0.40	0.40	0.40
Choline equivalents, mg/kg	1679	2257	2057	1485
(lb)	(763)	(1026)	(935)	(675)
Methionine (% of diet)	0.33	0.30	0.24	0.29
Cystine (% of diet)	0.32	0.28	0.18	0.22
Lysine (% of diet)	1.13	0.88	0.52	0.74
Tryptophan (% of diet)	0.29	0.26	0.20	0.21
Threonine (% of diet)	0.85	0.63	0.47	0.63

[a] Source: Summers and Leeson (1976).

In earlier days, goose fat had many uses around the home. Geese were force-fed or noodled to increase the amount of body fat as well as to increase the size of their livers, a principal ingredient in paté de foie gras. Noodled geese can gain 7–8 lb (3.0–3.5 kg) in 3–5 weeks and the liver size quadruples. The name noodling is derived from the moistened ground corn and wheat middlings that were rolled into cylinders about 2 in. long and fed to the geese four or more times per day. Noodled geese are produced only in isolated localities today and in small numbers because consumers prefer a leaner carcass.

Goose Diseases

Geese are relatively resistant to disease. Fowl cholera can be a problem in geese. Ophthalmia or sore eyes can be a problem in late summer and is usually associated with diets low in vitamins A and C. Dry weather and limited grazing materials may contribute or be the cause of the problem. Avian tuberculosis, caused by *Mycobacterium avium,* can be found in older geese. This slow-growing bacterium is susceptible to streptomycin. Leucocytozoonosis, caused by *Leucocytozoon semondi,* can also occur in ducks and geese. Black flies in summer that bite young unfeathered goslings and infect them are the vector. No suitable treatment exists.

PROCESSING WATERFOWL

Ducks and geese are processed similarly. They are shackled by the feet, stunned, and killed by severing the jugular vein which is done by inserting a narrow knife through the rear of the roof of the mouth and then into the brain just back of the eyes. Commercially, waterfowl are semiscalded in 140–160°F (60–72°C) water for 2–3 min. Waterfowl are mechanically defeathered usually in batch-type pickers. They are then dipped in molten wax one or more times, then in cold water to set the wax. When the wax is stripped off, the pinfeathers are removed with the wax. The feathers of commercially processed waterfowl are washed, dried, and sorted for sale to manufacturers of sporting clothes and equipment.

The dressing percentages for ducks and geese are lower than for chickens and turkeys. Dressing percentages of 67–75% including giblets are normal. In the United States most ducklings and geese are sold frozen.

How to hold the duckling or gosling

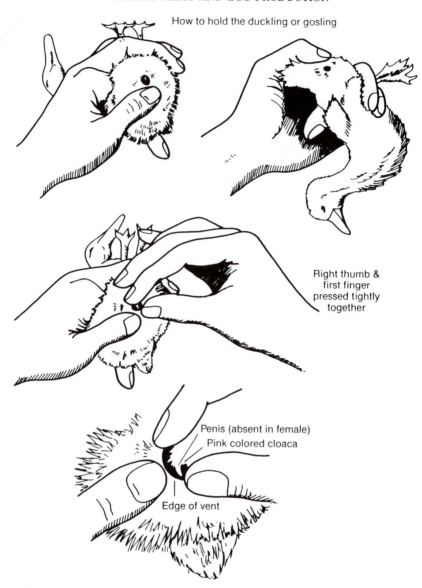

Right thumb &
first finger
pressed tightly
together

Penis (absent in female)
Pink colored cloaca

Edge of vent

FIG. 14.8.   Vent sexing waterfowl at hatching.

HANDLING WATERFOWL

Waterfowl should be handled carefully when being moved, weighed, or assessed as potential breeders. It is proper to grasp waterfowl by the neck rather than by the feet as is done with chickens. Waterfowl have short legs and can be injured by picking them up by the feet because this places stress on the hip joint and pelvis. Waterfowl can be carried by tucking the head and neck under one's arm and holding the feet loosely while letting the body of the bird rest on the forearm.

PINIONING

Some waterfowl are pinioned to make them flightless so they will not leave the area. Waterfowl can be temporarily pinioned by cutting off the primary feathers on one wing, upsetting the balance and rendering the bird flightless. Permanent pinioning can be accomplished by removing the last wing joint (metacarpus-phalanges) when the bird is young and cauterizing it. This operation prevents the growth of the primary feathers needed for flight.

SEXING WATERFOWL

Newly hatched waterfowl can be vent sexed by everting the cloaca and observing the presence or absence of a penis. Male waterfowl at hatching have a flaccid light-colored penis (see Fig. 14.8).

Adult waterfowl, particularly geese and swan, can also be vent sexed. Adults are laid on their back. The sphincter muscle around the vent is then stretched causing the cloaca to evert. When pressure is placed by the fingers above and below the vent the penis will protrude on the lower left side. The penis is in a sheath and is small in young ganders and 2–3 in. in length in adults. The opening in the functional left oviduct can be seen in females. The author prefers to hold waterfowl upright and hold them between his knees to sex them, a technique worth trying by the novice since large waterfowl never seem to appreciate the opportunity to lay on their backs to be vent sexed.

REFERENCES

American Poultry Association, Inc. 1983. The American Standard of Perfection, American Poultry Association, Inc., Troy, NY.

Ash, W. 1976. Raising Ducks. United States Department of Agriculture Farmer's Bulletin 2215.

.. Modern Waterfowl Management and Breeding Guide. American Bantam , Augusta, NJ.

.. S., and Aitken, J. R. 1961. Raising Geese. Canada Department of Agricul-.re, Ottawa, Ontario.

.w, D. 1975. Successful Duck and Goose Raising. Stromberg Publishing Co., Pine River, MN.

.ummers, J. D., and Leeson, S. 1976. Poultry Nutrition Handbook. Ministry of Agriculture and Food, Guelph, Ontario, Canada.

Miscellaneous Poultry

Poultry species that contribute limited amounts of income to total poultry sales are often referred to as miscellaneous poultry. These species are important even though total sales may be small. The less visible species of poultry include *Coturnix* quail, Bobwhite quail, Chukar partridge, ring-necked pheasants, guinea fowl, pigeons, and swan.

COTURNIX QUAIL

Coturnix coturnix is found in Europe, Africa, and Asia where it is considered a migratory species. Among the several recognized interbreeding subspecies are the European, *Coturnix coturnix coturnix,* and the Japanese, *Coturnix coturnix japonica.* In Japan, a semi-domesticated strain has been reared for pets, because of their singing ability, and for meat and eggs. Attempts have been made to introduce these subspecies as gamebirds in the United States but their introduction has been unsuccessful because the birds migrate. However, their attempted introduction brought them to the attention of breeders for their laboratory and commercial potential.

The subspecies, *Coturnix coturnix japonica,* has become important as a laboratory animal. Currently they are used for studies in embryology, endocrinology, genetics, nutrition, physiology, and toxicology. They have several characteristics that make them valuable for investigative purposes. They do well in laboratory cages and are inexpensive to maintain. Egg production is excellent with up to 300 eggs produced per year per female. The incubation period is only 17 days and the birds have a short life cycle, about 6 weeks from hatching to sexual maturity.

The advantages and disadvantages of *Coturnix* quail compared to chickens as laboratory animals are as follows.

Advantages

1. Coturnix occupy only 200 cm^2 of space.
2. They consume only 14 g of feed a day.
3. They reach maturity in 42 days.
4. They appear to be more disease resistant.
5. They respond favorably after surgery.
6. They are more photosensitive.
7. They have a higher metabolic rate.
8. Physiological aging is quicker because their life span is shorter than chickens.
9. They produce a greater volume of eggs per unit of weight and in a shorter time.

Disadvantages

1. *Coturnix* have an objectionable odor but it can be reduced by proper ventilation.
2. They waste feed by excess billage.
3. The colored eggshells make candling difficult.
4. They are easily excited.
5. The young birds have a poor homeothermic mechanism at hatching time and so are more susceptible to drafts and damp conditions.
6. Fertility and hatchability are variable.

Housing and Equipment

Since *Coturnix* are used extensively for research, housing, cages, and equipment should be designed to permit easy effective sanitization. They can be kept in commercially manufactured or custom designed cages. Feeding and watering equipment should be selected to make the feed and water easily accessible and the cages should be comfortable. The ventilation system should be of sufficient capacity so that there is enough exchange of air in the rooms or houses to prevent the accumulation of odors. Thermostats should be used to maintain room temperatures at between 50 and 100°F (10–38°C). Humidity should be kept within a range of 45–70%. The space requirements for *Coturnix* are shown in Table 15.1.

Waterers used for *Coturnix* should be narrow lipped to prevent the chicks from getting into them and chilling or drowning. If screw-on poultry drinking founts are used marbles or clean small gravel can be placed in the water to keep the water levels shallow enough to prevent drowning. Flat paper or egg flats can be used as feeders for the first few days. Later, feeders 2 in. high by 2 in. wide (5 × 5 cm) of any length can be used for larger *Coturnix*. A grill of welded wire laced in the feeder on top of the feed helps to reduce waste.

Breeders

Because of the characteristics of these birds, several areas deserve particular attention. Early maturing *Coturnix* can be mated in pairs or colony-type cages. If adequate housing space is available without lowering fertility, several females can be mated per male. Cannibalism and pecking in breeders are reduced if matings are made prior to

TABLE 15.1. Space Requirements for *Coturnix*[a]

Type of equipment	Age of bird (weeks)	Minimum space per 100 birds	Comments
Battery brooder	0–4	8361 cm^2 (9 ft^2)	
Colony cages	Over 4	14 cages; each 30.5 × 30.5 ×20.3 cm (12 × 12 × 8 in.)	Highest at rear sloping 15.2 cm (6 in.) in front; each is suitable for two males and five females; 1.25 m^2
Individual cages	Over 4	12.7 cm (5 in.) wide, 15.3 cm (6 in.) deep, 16.6 cm (6$\frac{1}{2}$ in.) high, sloping to 12.7 cm (5 in.) in front per bird	
Feeders	0–3	254 linear cm (100 in.), 4 chick feeders—30.5 cm (12 in.)	Less, if circular feeder is used
	Over 3	381 linear cm (150 in.)	
Waterers	0–3	2 pint-size Mason jar founts (1 pint = .47 liter)	Use special quail narrow-lip founts or use ring of $\frac{1}{4}$-in. hardware cloth to prevent wading
	Over 3	63.5 linear cm (25 in.)	Less, if circular waterer is used

[a] Source: National Academy of Sciences (1969).

sexual maturity. Most eggs or birds are from populations descended from one original source, so care should be taken not to increase the level of inbreeding. *Coturnix* are relatively healthy birds but several diseases are transmitted congenitally. These include *Salmonella* spp., *Mycoplasma* sp., infectious bronchitis, Newcastle disease, lymphoid leukosis, avian encephalomyelitis, and the adenoviruses GAL, CELO, and COFAL.

Nutrition

Young *Coturnix* need a high-protein diet similar to those commercially available as turkey starter or gamebird starter feeds, which have a protein content of around 28%. Layer diets need to be fortified with additional calcium and phosphorus because the high rate of laying of the quail hens depletes these mineral resources.

Egg Handling

Coturnix eggs weigh roughly 10 g. Most strains lay eggs with heavily pigmented shells. Eggs for hatching should be collected frequently and stored at 50–60°F (14 ± 3°C) and 60–80% relative humidity. Hatchability declines rapidly after a storage period of 7 days. For longer storage periods, eggs can be kept in polyvinylidene chloride bags for periods of 13–21 days with higher hatchability than unpackaged eggs.

Incubation

Quail eggs, like other eggs, should be incubated large end up and turned 6–12 times every 24 hr. In forced-air incubators a temperature of 99°F (37.5 ± 0.3°C) with at least 60% relative humidity should be maintained.

Brooding

Coturnix chicks require a high initial brooding temperature but the temperature can be decreased according to the following schedule (National Academy of Sciences):

| Age in weeks | Temperature at bird level | |
	°C	°F
0–1	38	100
1–2	35	95
2–3	30	85
3–4	24	75

Fewer brooding problems will be encountered if room temperatures between 60 and 70°F (16–22°C) can be maintained. Humidity can range from 30 to 80%. Although battery brooders work best for *Coturnix* they can be brooded on the floor.

Because *Coturnix* are often kept on wire, feather picking, cannibalism, and other vices frequently occur. Preventive measures include debeaking as early as 2 weeks of age and reduction of light intensity to 0.5 ft candle (5 lux). Some success in controlling cannibalism has also been achieved by increasing dietary fiber and arginine levels. In addition to the above measures, care should be taken to prevent overcrowding and overheating.

BOBWHITE QUAIL

In the United States the native Bobwhite quail, *Colinus virginianus,* is being reared in increasing numbers as a gamebird as well as for meat purposes. Since Bobwhite quail are native species, a state permit, which can usually be acquired through state wildlife organizations, is required to keep these birds in confinement.

Bobwhite quail can be identified by their characteristic call of "Bobwhite" of the male and the sex differences in coloring. They are larger than *Coturnix* quail but considerably smaller than Chukar partridges. Adult Bobwhites generally weigh in the range of 225–325 g depending on whether the strain is a "flying" strain or a more highly selected "meat-type" strain for processing.

Because Bobwhites are gamebirds, strong flyers, and much less domesticated than other commercial poultry species, care must be taken to house them in a manner that minimizes injuries from flying into the tops, sides, and ends of their pens.

Breeders

A Bobwhite colony can be started by securing adult birds, young chicks, or hatching eggs. Care should be taken to ensure that they are from a healthy disease-free source.

Bobwhites can be group mated or mated in pairs. Mating two females per male in cages normally gives good fertility and high egg production. Compared to *Coturnix* quail, Bobwhites mature slowly, reaching sexual maturity in 16 to 18 weeks of age when lighting is used. Egg production varies depending on management, nutrition, health, and vigor. Because they lay white eggs, they cover their nests in the wild rather than use pigmentation to camouflage them. The incubation period for their eggs is 23 days. Table 15.2 lists the approximate number of eggs produced per quail hen under several conditions.

TABLE 15.2. Production Guide for Bobwhite Quail Hens[a]

	Number of eggs
Normal mating season (no aritificial light)	50–100
Normal mating season (aritificial light 17 hr[b])	70–150
Year round production (17 hr of light)	150–200

[a] NCSU—Extension (see West 1987).
[b] Artificial lights never decrease the total hours of light during the laying period.

Egg storage, incubation, and brooding are similar to that used for *Coturnix* quail. Since Bobwhites are larger than *Coturnix,* more space is allocated during brooding and growing. Table 15.3 lists the space, feeder, and waterer requirements for rearing Bobwhite quail.

Nutrition

Bobwhites also require a high protein level in the starter diet of about 28%. A commercial turkey or gamebird starter generally meets these requirements. After 3 weeks of age the protein content can be decreased to 21–22%. After 6 weeks of age the Bobwhites should be fed according to whether they will be utilized as breeders, flight conditioned for shooting preserves, or processed for meat purposes. A major concern of many quail growers is controlling feed waste. Table 15.4 lists feed consumption for Bobwhites.

CHUKAR PARTRIDGE

The Chukar partridge *(Alectors chukar)* is a gamebird that appears to have some commercial potential. They originated in Asia and southern Europe. Stocks currently used in the United States come from the southern slopes of the Himalaya mountains. Chukars generally weigh about 16–24 oz (400–600 g), which is considerably larger in size than

TABLE 15.3. Space, Feeder, and Waterer Requirements[a]

Spacing	Age of quail		
	1–10 days	10 days–6 weeks	6 weeks–maturity
Floor	9 birds/ft^2	6 birds/ft^2	3 birds/ft^2
Feed	$\frac{1}{2}$-in./bird	1 in. bird	1$\frac{1}{2}$-in./bird
Water	Three 1-gal fountains/100 birds	1 linear ft/100 birds	1 linear ft/100 birds

[a] Source: NCSU—Poultry Extension (see West 1987).

TABLE 15.4. Bobwhite Feed Consumption Guide

Age	Approximate feed consumption (lb/bird)[a]
1 day–8 weeks	1.3–1.5
8–16 weeks	2.5–3.0

[a] Feed usage above these levels indicates feed wastage. Check feeder height adjustment and consider placing hardware cloth strips in feeders to prevent excess billage and scratching of feed.

quail. For this reason, Chukars have a greater potential for the restaurant trade, since an individual bird is just the right size to be used for an individual portion.

Chukars have red feet, legs, and beaks. Both the male and female have similar plumage, but male chukars can be identified by the spur buds that develop when they mature. Chukars are raised both for shooting preserves and for processing for consumption as a specialty meat item.

Breeders

Chukars are monogamous in the wild but they can be mated in pairs or trios in cages. Breeders kept in floor pens can be mated in a ratio of one male to four or five females. Chukars can be light stimulated and are better egg producers than pheasants and less flighty than Bobwhites or pheasants. Chukars mature more slowly than quail or pheasants.

Incubation and Brooding

Chukars lay a speckled egg that has an incubation period of 23 days. Young chukars can be fed and brooded the same as quail and pheasants. They are easier to raise than Bobwhites or pheasants because they are calm and seldom prone to feather picking or cannibalism. Like most gamebirds, Chukars grow slowly compared to broiler chickens. They are generally 12 to 16 weeks old before they are ready for processing.

PHEASANTS

Pheasants originated in China and have been transplanted throughout much of the world. There are many types of pheasants but only the ring-necked pheasant *(Phasianus colchicus)* and some closely related varieties developed in captivity have commercial potential. Pheasants owe a great deal of their popularity to the beauty of the plumage of the cocks. They have been successfully transplanted to the United States since the 1880s when they were first released in the Willamette Valley in Oregon. Other successful releases spread the range of the pheasant across the United States and as far south as Virginia, but they have not been successfully transplanted to the southeast. Stocking attempts that failed in the south were attributed to less suitable agricultural crops, low limestone content of the soils, and the wide use of pesticides in the traditional cotton and tobacco crops.

Breeder Pheasants

Pheasants reach sexual maturity at about the same age as chickens, 20 to 24 weeks. Under natural conditions, pheasants reach this age in early fall when day length is decreasing so they do not lay until spring when day length increases again. Since pheasants are closely related to chickens, most of the management techniques used with chickens can be applied to pheasants. For example, pheasants can be light stimulated and brought into egg production throughout the year if necessary. Most pheasants are mated at a ratio of one cock to five or six hens and kept in floor pens or outside run breeding pens. Pheasant hens can be caged in layer cages and artificially inseminated. For ease of insemination, the tail feathers of the hens are clipped before caging as the tail feathers hinder movement in cages. This method requires more expertise in artificial insemination and semen collection than most producers possess.

Pheasants are aggressive, slightly territorial birds and they can be cannibalistic. Birds should be debeaked frequently to help prevent cannibalism. Some producers with floor pen-type breeding facilities have had success in controlling cannibalism by making the pens three dimensional to allow dominated birds to avoid face-to-face contact with dominant birds.

Egg Production and Care

Pheasants are ground nesters. The pheasant hen makes a shallow nest in the ground or litter and lays her eggs. Their eggs are heavily pigmented and are olive to greenish-brown in color. Eggs should be gathered frequently to prevent them from becoming broken or soiled. Egg eating can become a problem in breeder pens if the eggs are left in the pens. The number of eggs produced per hen varies depending on the housing, nutrition, lighting schedule, and management. Hens can lay from 50 to 125 eggs or more during a year. Their eggs, which have an incubation period of 23 days, should be incubated the same as chicken eggs.

Brooding and Rearing

Newly hatched pheasant chicks can be battery or floor brooded. Because pheasant chicks are adept at escaping from their pens care should be taken to keep them confined close to the heat source, feed, and water. Pheasants, like young Bobwhite quail, learn to fly at an early age. The tops of battery brooders should not be over 8–10 in. (20–25 cm) high to prevent injuries to the backs and necks of birds flying into the tops of the brooders. Pheasants can be fed a gamebird feed or turkey starter initially since their protein requirements are similar to those of quail and turkey poults. Young pheasants are prone to cannibalism and should be debeaked for the first time at about 2 weeks of age. Further beak trimming may be needed at 4–6 week intervals. Overcrowding and overheating should be avoided to help prevent cannibalism.

Commercially grown pheasants are generally marketed for shooting preserves or for meat. Shooting preserves are a major outlet, particularly for rooster pheasants. Others are commercially processed and sold frozen to the restaurant trade and specialty meat markets.

GUINEA FOWL

Guinea fowl are one of the lesser known poultry species. Wild species are still found in Africa where they originated. "Guinea" is derived from the name given to a portion of the west coast of the African continent. The domestic stocks were derived from the wild species *Numida meleagris* that are found in that region.

Guineas appear to have been domesticated for almost as long as chickens. They were reared by the ancient Greeks and Romans. In the United States guineas were brought to the colonies by the early settlers. They are excellent foragers and can provide the majority of their own diet.

Guinea production in the United States is rather limited. It is estimated that approximately 5 million birds are produced annually. This production is scattered over the southeast, midwest, and northeastern states with many of the birds grown in small flocks. A few specialized farms grow guinea fowl on a commercial basis.

Guineas are delicious when they are processed at a young age and properly cooked. The young guineas are tender and have a rich gamelike flavor. They are often served at banquets and in fine restaurants for special occasions. They have a flavor similar to quail, partridges, and pheasants. In some areas they are marketed as African pheasant. Because of the plentiful supply of other poultry products, it appears that guineas will never reach the popularity that they have in France, where production is substantial because a demand exists for their game-like taste.

Breeding Guineas

Many inexperienced growers of guineas have trouble determining the sex of their birds because the males and females are similar in appearance. The male guineas, called jacks by many people, have a larger helmet and wattles and are generally coarser around the head than the female. The females, or jennies, are more refined and utter a two-syllable cry that sounds like "buckwheat, buckwheat," or "put-rock, put-rock" as described by some growers. The males or jacks use only a one-syllable shriek.

Wild guineas tend to mate in pairs, but domesticated guineas in confinement can be mated in a ratio of one male to six or eight females. They can also be reared in cages and artificially inseminated with good results.

Varieties of Guineas

In the United States there are three principal varieties of guineas: Pearl, White, and Lavender. The Pearl or wild color in guineas is a grayish-purple that is dotted or pearled in white. This variety is the most popular in small flock situations. Some of the feathers are used for ornaments. The Lavender variety is lighter than the Pearl variety in color and is often described as light gray or lavender or dotted with white. The White variety has white plumage and is favored by processors because of the more attractive carcasses and lighter skin.

Guinea breeders can be managed like breeder hens of other species. They respond to artificial lighting. They have a higher protein requirement than chicken breeders but respond well to a breeder diet of 22–24% protein. Egg production of guinea hens depends on breeding, management, and nutrition. Many guineas lay 50–100 eggs per year. Well-managed, confined guinea hens often lay in excess of 100 eggs per year. In farm flocks guinea hens often are kept 2 to 3 years or longer. Commercially, they are generally replaced after one laying cycle.

Guineas that are allowed to roam free often lay 25 to 30 eggs and then turn broody to incubate their eggs. Hens hide their nests because they often have problems with predators during the incubation period. The eggs weigh 16–18 oz per dozen and are smaller than chicken eggs. Guinea eggshells are harder and have fewer pores in the shell than chicken eggs.

Incubation

Guinea eggs have an incubation period of 26–28 days and can be incubated the same as turkey eggs. In small flocks, eggs can be set under either chicken or guinea hens for incubation. However, forced air incubators operating at 99.5–100.0°F and a relative humidity of 57–58% are used in commercial hatcheries. When guinea eggs are transferred to the hatcher, a humidity of 60% appears optimum.

Young guineas are called keets and are attractive chicks. Their beaks and shanks are red and keets from the Pearl and Lavender varieties also show the wild down pattern seen in many gamebirds.

Young keets can be commercially brooded the same as chicks or turkey poults. They can be quite excitable when several months old and behave much like floor-reared leghorn chickens. Commercially, keets can be reared on diets similar to those used to rear turkeys. Guinea fowl in small flocks are marketed at 15–16 weeks of age and weights of around 3 lb. Markets generally prefer dressed weights of at least 2 lb. To date, guinea fowl in the United States have had little commercial application, so only limited research and genetic selection pressure have been applied to them, but commercial hatcherymen and growers are now making progress in egg numbers, age to market, and body size, making guineas more competitive with other poultry. Some producers are now reporting keets of market size at 11 to 12 weeks of age.

PEAFOWL

Peafowl today are strictly an ornamental bird species. The males or peacocks have been famous in art and legend for centuries, not only in India and China, but around the Mediterranean sea as well. They have been important objects in religion and legend over time. Nearly every individual has at some time admired them in parks or zoos. Peacock feathers are widely used for decorative purposes.

Two species of peafowl exist. The most common of the two is the Indian peafowl, *Pavo cristatus*. This species of peafowl is native to India and Ceylon, but has been transported almost everywhere in the world in which other poultry is grown.

Breeding Peafowl

Peafowl mature slowly compared to other poultry species. The hens can start to lay at about a year of age but most start to lay the second year. The hens are ground nesters and normally lay four to eight eggs, with five or six being the average. If the hens are confined and their eggs removed, they will lay an additional clutch. The incubation period of 27–30 days is similar to turkeys. Peachicks can be brooded artificially or by the peahen. The chicks are able to perch and fly short distances at an early age. The males or peacocks do not develop full plumage until they are 3 years old. Contrary to what many people believe, the long feathers of the peacock's train are not tail feathers. The train is made up of 100–150 tail coverts that, when spread, are supported by the stiff rectrices. Young peacocks do not develop a train until they are 3 years old. After that time the train is molted annually in the late summer after the breeding season. Young peacocks are sexually mature by 2 years of age, but seldom are able to attract hens without the train for courtship or because they are dominated by older males. Peafowl can be mated at a ratio of one cock to four or five hens.

Nutrition and Feeding

Peafowl that are allowed to wander loose are good scavengers. Their diet can be supplemented with some grain or a complete poultry feed. Confined peafowl can be given a complete layer diet, preferably in pellet form.

Color Mutations in Indian Peafowl

The color of Indian peafowl is commonly referred to as Indian blue, but over time color variations have appeared. Albinism is frequent and white Indian peafowl are common. Pied peafowl have also been established in which patches of white replace colored areas. Another color variation is the Black-winged or Black-shouldered peafowl, a mutation that occurred in the 1800s. The scapulars, wing-coverts, and tertiaries are black instead of barred buff.

The other species of peafowl is the Green peafowl, *Pavo muticus,* which inhabits areas further east of the Indian peafowl in Burma, Thailand, Indochina, and Java. Green peafowl can be recognized by their longer legs, more erect stance, long narrow crest, and metallic green color. They can be reared similar to Indian peafowl, but they are less hardy and need shelter when temperatures drop below freezing. The voice of a Green peacock is less offensive than an Indian peacock but they are more quarrelsome and aggressive.

Indian peafowl and Green peafowl will cross and produce fertile hybrids. Black-shouldered peacocks were crossed with Green peahens to fix and develop the hybrid variety known as Spalding's peafowl.

Peafowl tend to have few diseases, but they are particularly susceptible to blackhead (histomoniasis). Because chickens can be the intermediate host for the blackhead organism, it is a good management practice to keep peafowl and chickens separate and not allow peafowl access for several years to areas in which chickens have been kept.

PIGEONS

Pigeons *(Columba livia)* appear to be descendents of the semiwild blue rock doves of Europe, Asia, and Africa. They have been domesticated for centuries, primarily for food production. However, breeds of pigeons have also been developed for racing, carrying messages, exhibition, and entertainment.

Pigeons differ in several major ways from the other poultry discussed. They have altryical young, meaning they are hatched blind, naked, and unable to feed themselves. Therefore, when pigeons hatch their young they are cared for by the parents. This characteristic also means that pigeons cannot be mass produced in large numbers as are chickens, turkeys, and waterfowl.

There are many breeds of pigeons, but only a few breeds are used for meat production. In producing pigeons for meat purposes only the young pigeons, called squabs, are used. Surprisingly squabs grow rapidly and are ready for processing by 28 days of age. Squabs are said to be ready to process when they feather under the wings. They are excellent meat and are to poultry meat what veal is to beef. Because squabs are slaughtered very young, they are tender and high in water content and the meat is very moist. Also, since they are processed before they leave the nest and therefore have never flown to exercise their muscles, they are very tender. Currently squabs in the United States sell for $3.50–5.00 each, wholesale, and are purchased by restaurants, country clubs, and hotels. Squabs are especially popular on the West coast, in the Chicago area, and in New York.

Breeder Pigeons

It is difficult for most persons to determine the sex and age of pigeons by casual observations. If the sex of several pigeons is unknown they can be penned together and observed; the male or cock pigeon will coo and court the hens. Pigeons reach sexual maturity between 6 and 8 months of age depending on the variety. They mate in pairs and tend to stay with their mate, but pairs can be broken up and remated. They can be used as breeders as long as they are productive, which can be from 3 to 7 years, but most breeders become less productive after 5 years of age.

Pigeons normally lay two eggs, skipping a day between eggs. Both the cock and the hen share the incubation of the eggs, which lasts 17 days, and the feeding of the young squabs. The young newly hatched squabs are fed by regurgitation by the parents. Their first food is often called "pigeon milk." Actually under the influence of rising prolactin levels that occur as the cock and the hens share incubating the eggs, the crop lining thickens with layers of thecal cells. When the parents regurgitate, these thecal cells are shed and fed to the squab. These cells are high in protein and ensure rapid growth. Pigeon milk levels drop after about a week and crop contents make up a larger portion of the diet. The squab are fed by the parents until they are slaughtered or leave the nest after they learn to fly and feed themselves. Squab-type pigeons need to raise 12–18 squab per pair per year to be profitable, since it has been estimated that 10 squab per year is the breakeven point. Production in other breeds and varieties of pigeons is much less.

Breeds of Pigeons

There are numerous breeds and varieties of pigeons. In the United States homing pigeons are popular; Fantails, English Carriers, and Runts are seen at shows and vary in price and quality. Squab pigeons in the United States are mainly White Kings or White Carneaux or color varieties of them. A pair of young squab type breeders costs $25.00 or more per pair in the United States.

Breeder Facilities

Breeder pigeons are normally mated in multipair housing units each holding 15–25 pairs of breeders. Care should be taken to ensure that they are matched pairs and no extra males are present to cause strife in the unit. A breeding unit of this capacity is generally 12–16 ft^2 (4–5 m) and has an additional screened-in sun porch or small flight pen attached.

Generally for squab production two wall-mounted nests are provided side by side for each pair. Breeders in top production will lay in the second nest before the squabs are ready to harvest at 28 days of age. Various types of equipment are available for pigeons or poultry equipment can be adapted for this use.

Covered tube-type hanging feeders mounted several inches off the floor work well as feeders. Because pigeons often contaminate their water with their droppings, which can lead to disease problems, several pigeon operations are using nipple waterers or small self-filling cups designed for caged layers.

Feeding

Pigeons can be fed a complete pelleted ration or mixture of a complete pellet and whole or cracked grains freely. Complete breeder rations are normally 20% protein. Since pigeons are often given whole or cracked grains grit is normally provided freely as well.

Management

Pigeons, when properly cared for, are relatively free of disease. When given proper housing and nutrition disease problems should be minimal if high levels of sanitation are maintained. Successful growers take precautions in bringing new stocks to the facilities, limit visitations, and birdproof pens to keep out sparrows, starlings, and free flying pigeons that are potential sources of disease and external parasites.

SWAN

Swan are the aristocrats of the waterfowl (Fig. 15.1). They are some of the largest, most long lived of the waterfowl, often living 30–50 years or more. Swan can be differentiated from geese by their longer necks and unfeathered lores and because they are less agile on land. Swan tend to mature slowly, with some breeds requiring 3–4 years to mature. Generally, both sexes assist in nest building. In swan the male is called a cob and the female is called a pen. Young and partially grown swan are called cygnets.

Breeds of Swan

The most popular of the swan kept in the United States is the large white Mute swan *(Cygnus olor)*, often called the Royal Mute swan. This breed is native to Europe. It was taken from the wild and now has become quite domesticated. Mute swan are kept for their aesthetic value and can be found scattered throughout the world. The cob (male) can be identified by the more prominent knob extending forward over the base of the bill.

Another popular breed of swan is the Black Australian Swan *(Cygnus atratus)*. These birds appear black in the distance but their plumage is really a dark grayish black with white primary feathers and dark feet and legs. Black Australian swan have a bill that is deep-orange salmon in color with a white band across the tip. They reproduce at a younger age than the Mute swan.

FIG. 15.1.   A pair of white Mute swans.

Breeding Swan

Swan should be vent sexed to ensure that they are really a pair. They are becoming increasingly popular in the United States with breeding pairs costing $500–750 per pair. Pairs of swan become quite territorial and aggressive during the breeding season. They form strong pair bonds and remain together as long as they both stay healthy. If something happens to one of a pair the other will find a new mate. One reason many pairs of swan do not reproduce is because they do not reach a point of reproductive fitness because of competition for feed. Ducks and geese feed more rapidly than swan so in ponds having mixed waterfowl, the swan seldom receive enough to eat. The problem can be alleviated by placing a feeder mounted in water on a post out of the reach of other waterfowl.

Incubation and Brooding

Swan only lay 4 to 6 eggs in a clutch and are usually allowed to incubate their own eggs. The eggs normally hatch in 34–40 days.

Young swan are carefully protected by the cob and pen. However, because cygnets can fall prey to large turtles and fish, they swim with their parents. The safest solution is either to rear the cygnets separately from the parents or pen the parents and cygnets off in shallow water free of predators. Young cygnets should be pinioned at an early age, preferably before 4 weeks of age. Cygnets should be removed from the breeding territory of the mated pair before the next breeding season or the cob will drive them from the area.

Mute and Black Australian swan are nonnative species in the United States so a holding permit issued by the U.S. Fish and Wildlife Service is not required. However, a permit is required for swan such as Whistling and Trumpeters. Because of their value, it is a good idea for swan owners to toe punch or tattoo the webs of the feet of their birds for positive identification.

REFERENCES

Cooper, J. B. 1944. Poultry for Home and Market. Turner E. Smith, Atlanta, Georgia.

Delecour, J. 1977. The Pheasants of the World, 2nd Edition. Spur Publications, Sarga Publ. Co., Ltd., Surrey, England.

National Academy of Sciences. 1969. *Coturnix*. Publication 1703, National Academy of Sciences, Washington, DC.

National Academy of Sciences. 1977. Laboratory Animal Management: Wild Birds. National Academy of Sciences, Washington, DC.

USDA. 1976. Raising Guineas. Leaflet 519, U.S. Government Printing Office, Washington, DC.

West, J. R. 1981. Bobwhite Quail Management. North Carolina Agr. Ext. Service, NCSU, Raleigh, NC.

Winter, A. R., and Funk, E. M. 1941. Poultry Science and Practice. J. B. Lippincott, New York.

16

Poultry Management Practices

A number of poultry management practices have been developed and adopted by the various segments of the poultry industry. Although a few have been adopted universally and become standard practice in the industry, most practices still depend on the producer's situation and management of his flocks. Other management practices are covered in Chapters 11, 12, 14, and 15.

PRODUCTION MANAGEMENT

Types of Poultry Feeders

To encourage poultry to learn to eat early in life, the first feed is usually spread over a large flat container called a feeder lid or placed on paper for easy access by the young poultry. These devices are removed by the end of the first week and the permanent feeders are utilized thereafter. There are several types of feeders, some automated and others hand filled. Hand-filled feeders are often used in developing countries in which labor is cheap. In the United States, they are utilized in special situations in which feed restriction is used or some similar circumstance, but automatic feeders are used almost entirely in commercial scale operations.

Hand feeders can be of two types—trough or tube feeders. Troughs used in the past have been 4–6 ft long (1.2–1.8 m) covered by a grill to prevent the birds from getting into the trough. Tube feeders can be constructed of sheet metal or plastic in 12- to 16-in.-diameter (30–40 cm) cylinders, about 2 ft (0.6 m) long attached to an adjustable suspended pan. These feeders are generally suspended from the ceiling and have a capacity of 30–50 lb (13–22 kg).

Automatic feeders are available in several types. Generally, they consist of a pan or trough from which the birds eat, and a mechanical filling system that brings feed from the feed storage tank.

Popular automatic feeders include the conveyor-and-pan type which uses a chain or auger to push feed through a tube dropping it into pans. Sensors stop the flow of feed once the pans are filled. Tube and trough systems move feed along a tube that has a trough attached to it with holes in the tube to allow feed to go into the trough. Hanging cylindrical tube feeders utilize an auger or cable with disks to move feed through the tube. Drop tubes at intervals permit feed to drop down and fill the feeders.

Types of Poultry Waterers

Poultry waterers, like feeders, can be hand filled or automatic (Fig. 16.1). When starting young poultry, 1-gal. (3.78 liters) waterers can be used for the first 5 days, the same as feeder lids are used, prior to the use of permanent feeders. Trough-type waterers are generally not acceptable for newly hatched poultry but work well for growing turkeys, pullets, and some breeders. Valved-cup waterers are marketed for poultry of all ages. Drip-type waterers are also available; they allow the bird to pick off the droplets and catch them as they gradually flow from the drinkers (Fig. 16.2). Caged layers utilize drinking cups very well. Most poultry producers in the United States utilize the bell-shaped, low-pressure plastic drinkers that are suspended and contain an internal ballast to prevent spillage (Fig. 16.3).

Additional watering equipment should include a water meter, a proportioner, and an auxiliary backup watering system. Proportioners are mechanical devices that include a pumping mechanism that meters vaccines or medicinal stock solutions directly into the waterline. They generally meter 1 oz (30 ml) of medicine into every gallon of water (3.78 liters). Table 16.1 gives an estimate of water and feed space requirements. Table 16.2 shows the approximate water consumption of poultry of various ages.

FIG. 16.1.   Automatic chick waterers designed for use particularly during the brooding period.

Litter

Almost a hundred different materials have been used for litter with varying degrees of success. To be usable, litter must be inexpensive and readily available, free of dust, dry, and free of molds. It should also be small enough and heavy enough so that feed hoppers are not contaminated excessively with the litter. In addition to having a high moisture absorbing capacity, it should be comfortable when in contact with birds. It should not be cool to the touch, which is the case with mineral-type litters.

Wood shavings, sawdust, and peanut hulls are the most popular types of litter material. Other materials used for litter with varying degrees of success include sand, shredded newspaper, shredded sugar cane, rice hulls, ground corn cobs, and several types of mineral materials. Generally, litter is used for several flocks of birds. Litter material

FIG. 16.2. Drip waterers allow the bird to pick off the droplets and catch them as they flow from the drinkers.

containing the dried droppings can be reconditioned to make a satis-factory litter for several successive broods. Before a new flock is started the litter can be aerated when necessary. Wet and matted litter is removed and replaced with fresh material. With proper management, litter can be used to help keep environmental conditions in the house in balance by helping to prevent extremes of humidity and tempera-ture, as well helping to prevent high levels of gases, particularly ammonia.

Brooders

Brooders may be classified on the basis of their fuel source, electric, gas, oil, coal, wood, and solar, and on the basis of the many variations in design for the above types of fuel. The most commonly used brooders for commercial production are gas and electric. Other types have been generally discarded because of their high labor requirements and lack of flexibility in maintaining uniform temperature ranges throughout the year.

Brooding systems can be classified further on the basis of central heating systems and individual units. In central heating, gas, oil, or coal is used as the heat source. The heat is distributed through hot air ducts or hot water pipes generally with fans to improve heat distribu-

FIG. 16.3. A flock of young broiler chickens. Note the problem of chicks in the feeders and the bell-shaped, low-pressure, suspended, plastic drinkers.

tion. A sheet metal covering or canopy is usually used to keep the heat at floor level.

Brooders can also be classified as warm room brooders and cold room brooders. Electric and gas heaters provide cold room brooding. The heat generated radiates directly down on the floor area under the canopy, but provides little additional heat for the rest of the house.

Warm room brooding can be provided by gas and by combinations of several sources to provide heat at reasonable prices. Although canopies

TABLE 16.1.   Feeder and Waterer Space Requirements for Chickens and Market Turkeys

Age	Feeder space/bird		Waterer space/bird	
	in.	cm	in.	cm
Through 2 weeks	1	2.5	⅓	0.8
3–6 weeks	2	5.0	⅔	1.6
7–18 weeks	3	7.5	1	2.5
19 and older	4	10.0	1	2.5

TABLE 16.2. Water Consumption[a] by Chickens and Turkeys of Different Ages[b]

Age (weeks)	1000 chicken broilers/day		1000 egg strain pullets/day		1000 turkeys/day	
	Liters	Gallons	Liters	Gallons	Liters	Gallons
1	23	6[c]	19	5[c]	37	10[c]
2	42	12[c]	38	10[c]	75	20[c]
3	67	17	45	12	113	30
4	126	34	64	17	151	40
5	140	38	83	22	189	50
6	170	47	94	25	227	60
7	207	56	105	28	283	75
8	235	64	113	30	359	95
9			132	35	434	115
10			143	38	473	125
12			151	40	567	150
15			158	42	605	160
20			170	45		
35	Laying or breeding		190	50	700	185
					450	119

[a] Will vary considerably depending on temperature.
[b] Source: Ryan (1978).
[c] By end of week.

are used, the heat radiates out from under the canopies and heats the entire room. Warm room brooding is becoming increasingly popular as newer methods of brooding, such as partial house brooding at high density, become popular.

SPECIALIZED PROCEDURES AND DEVICES

Beak Trimming

Birds in confinement often become cannibalistic and severely injure or even kill other birds. Beak trimming has been demonstrated to be an effective method for controlling cannibalism. For this reason, many breeders and growers employ the practice routinely. The beaks are usually trimmed at hatching time. Pullets for laying flocks are generally retrimmed or "touched up" a second time when they are between 12 and 17 weeks of age. Improper beak trimming or beak trimming pullets after 17 weeks of age can cause problems and slow down growth and development.

Beak trimming is done in several ways. One popular method is to use an electric shear-type debeaking instrument. The end of the beak is not only sheared but cauterized by the hot blade to prevent bleeding. In the past, either one-third of both the upper and lower mandible was cut off, two-thirds of the upper mandible only was cut off, or one-third of the upper mandible and just the tip of the bottom mandible was removed. Newer methods of debeaking literally burn the upper mandible. As the chick grows, atrophy occurs and blunts the upper mandible. This method is faster and has the advantage that the mandible is intact, initially, and the bird can eat normally. A laser-type debeaker that arcs between ceramic cones is also currently in use. It burns a small hole through the upper mandible and then atrophies and blunts the beak. Debeaking should be done carefully and thoroughly to prevent injury, discomfort, and retardation of growth and development of the bird.

Desnooding

Desnooding is the removal of the fleshy appendage on the turkey's head just behind the base of the upper beak. Removing snoods of male poults supposedly reduces fighting, prevents head injuries, and may prevent or reduce the incidence of erysipelas. Male poults are usually desnooded at the hatchery when requested by the purchaser.

Toe Clipping

Toe clipping, or removal of turkeys' toenails, is done at the hatchery. Turkeys when crowded or when being loaded for transport often step on each other and cause tears on backs and sides that reduce the grade of such birds at market time. Either surgical scissors or a modified hot-blade debeaker can be used for toe clipping.

Dubbing

Dubbing is the practice of cutting off the combs of baby chicks or, less frequently, of older birds to prevent frostbite, damage from fighting, or catching their combs in the wires of laying cages when eating or drinking. Commercial broiler breeder males are normally dubbed.

Banding

Wing banding, in which a number punched on an aluminum loop is inserted through the web of the wing of chicks at hatching time, is used to identify birds. It is used mainly in breeding programs.

Leg Banding

Leg banding, consisting of a number punched on an aluminum loop, is also used to identify birds. Leg bands are used on breeder birds and to identify birds which were blood tested.

Pick Guards

Pick guards or hen specs are aluminum or plastic devices that fit over the top beak. They are fastened through the nostrils of the chicken to prevent the bird from picking other birds.

Devices for Restricting Flight

On occasion it is desirable to either permanently or temporarily restrict the flight of birds, particularly gamebirds, so they can be caught and handled more easily and to prevent the birds from injuring themselves. This can be done in several ways.

Clipping the first 10 primary feathers with heavy scissors is the easiest way to temporarily restrict flight. As soon as the bird molts, new primary feathers grow out.

Brailing is another method for temporarily restricting flight. A brail, usually of leather, is attached to one wing so the bird cannot fully extend the wing to fly. The brail can be adjusted from time to time but to avoid permanent loss of flight, it is necessary to change it from one wing to another three or four times a year.

Pinioning permanently eliminates a bird's ability to fly. In pinioning the outer section of the wing is cut off when the birds are 4 to 7 days old. Generally, the wound is cauterized with tannic acid or heat to prevent bleeding. The operation is more traumatic on older birds because it is often necessary to tie and keep a cord on the second joint of the wing for several days to prevent excessive bleeding. The problem with this method is that it requires additional handling to remove the cord.

Artificial Insemination

Artificial insemination has become a common practice in the turkey industry and to some extent it is used by chicken breeders. To collect semen, the male is generally held with the keel in the palm of the hand with the legs held by the fingers. Then the back is stroked from midpoint toward the tail with the right hand, and the abdomen is massaged by the fingers holding the bird. After several vigorous strokes the thumb and forefinger of the free hand are used to apply pressure to both sides of the vent. Pressure is applied to the abdomen at the same time. This procedure should cause a flow of whitish semen from the vent. The semen is then caught in a suitable container.

Only females in production should be inseminated and never when a hard-shelled egg is in the lower portion of the oviduct. Hens are held and stimulated the same as males; an orifice or opening will appear on the left side of the vent when sufficient pressure is applied to the abdomen. A 1-ml syringe or medicine dropper containing the semen should be inserted $\frac{1}{4}$ to 1 in. (1–2 cm) into the orifice and the semen evacuated from the syringe. Normally, $\frac{1}{40}$ to $\frac{1}{20}$ ml (0.025–0.05 ml) of semen is sufficient per insemination. One insemination every 7 to 10 days should ensure high fertility.

Sexing

Sexing of chicks and poults is done for turkeys and recently for some chicken broiler flocks so the sexes can be housed and grown separately. Chicks for laying flocks are also sexed so that the cockerels can be discarded. Sexing is a job that requires highly trained professional sexors who charge a fixed fee per bird.

Methods of sexing include cloaca eversion, feather sexing, and color sexing. Cloaca eversion involves examination of the genital area of the vent. Feather sexing uses rapid and slow feathering traits, which can be fixed in the breeding program to make feather sexing possible. Color sexing uses sex-linked color traits to develop color sexing offspring by using the silver and gold gene.

Caponization

The caponizing (castration) of male chickens dates back to the pre-Christian era. Young chicken males are caponized to improve the quality of the meat. Today, any of the young males from broiler strains make good capons. In the past, Barred Plymouth Rocks or similar dual-purpose breeds were often used for capon production. Strong, healthy, large-framed cockerels should be selected to caponize. The operation is most successful on cockerels weighing less than 2.2 lb (1 kg). Fasting (feed and water restriction) for at least 12 hr allows the gut to empty, improves visibility during surgery, and is less stressful on the bird.

Surgical Procedure for Caponization

For caponization, the cockerel should be restrained by placing it on its side and fastening its legs and wings with enough tension to stretch out the bird. After removing a few feathers from the incision area and locating the last two ribs, the skin should be pushed up and back toward the thigh. Care should be taken to ensure that the thigh muscle is out of the way while making about a 1-in. (2.5 cm) incision through the skin and tissue between the last two ribs. After the rib spreaders are inserted, the peritoneal membranes and abdominal air sacs are gently torn to expose the testicles. The lower testis is removed first. Care should be taken to remove the entire testis so that regeneration will not occur. A cockerel incompletely castrated is called a "slip." For proper removal, the testicles should be drawn out with a slight twisting action. Care must be taken to prevent the rupture of any major blood vessels to prevent internal hemorrhaging, which can result in death. When both testicles have been removed, the rib spreaders are removed. When the cockerel is released and stands up, the skin and thigh muscles cover the incision. Since chickens have a high body temperature there is little chance that infection will result from the less than aseptic procedures utilized.

Capons do not require special care but wind puffs will develop in some birds from air escaping from the body cavity through the incision. These puffs can be punctured until the incision heals. Capons are less popular than they once were because of the development of rapid growing roaster lines that produce roasters with qualities similar to those of capons.

Judging

Poultry judging, in addition to providing opportunities to learn about live birds and the basis of grade and quality of poultry products, helps students learn to participate in competitive events and defend decisions. Since fewer and fewer individuals produce poultry, less emphasis is placed on poultry judging. In the United States today, however, many youth belonging to the 4-H and the Future Farmers of America (FFA) still compete in local, state, and national judging contests. At the collegiate level, the University of Arkansas hosts a National Collegiate Poultry Judging Contest and the University of Tennessee hosts a Southern Collegiate Poultry Judging Contest. The value of these contests varies. The exposure of young people to our modern industry is invaluable in itself. Other benefits include potential career options, learning experiences, competition, and the value of sound decision making which is useful in everyday living.

Generally, contests involve placing live poultry and grading poultry products according to USDA Guidelines. This participation allows students to learn about industry standards for live poultry as well as USDA standards for products.

Showing Poultry

The breeding, rearing, and showing of poultry for exhibition still has many followers. Since commercial poultry emphasis is now on utility most commercial poultry are now crossbred. However, the fundamental characteristics of exhibition poultry should be preserved for the following reasons.

The current genetic pool of poultry stocks may be needed in the future.

Exhibition poultry keeping is an excellent hobby and provides learning and social experiences for the participants.

Exhibition poultry can be utilized as a teaching tool for young people, as well as for teaching responsibility. Hopefully, through various youth activities more young people will develop career interests in working with poultry.

SANITATION AND WASTE DISPOSAL PROCEDURES

Poultry Waste Handling and Utilization

Poultry manure consists of both the feces and the secretion of the kidneys which contain the nitrogen in the form of solid uric acid. Much of the fertilizing value of the nitrogen in poultry waste is lost when putrefaction occurs and the uric acid changes to ammonium carbonate.

In modern commercial poultry operations, there are two types of poultry waste: feces or manure from caged layers and feces from floor-reared turkeys and chickens mixed with litter materials. Fresh poultry manure is high in moisture, around 75% and is about 20% organic matter. Table 16.3 gives the manure production rates and approximate fertilizer content of various poultry types.

TABLE 16.3. Manure Production Rates and Approximate Fertilizer Content on a Manure Weight Basis[a]

| | | Fresh feces and urine per animal | | |
| | | | lb/ton of fresh waste | |
	Animal weight unit (lb)	Tons/year	Nitrogen (N)[b]	Phosphorus (P)	Potassium (K)
Caged layer	4	0.05	26.7	9.0	9.6
Broiler[c]	4	0.01	52.7	18.0	27.5
Turkey	15	0.10	30.0	9.7	9.2

[a] Source: North Carolina Agr. Extension Service (1973).
[b] Generally, about 50% of nitrogen in fresh waste is volatilized and not available for fertilization, so only about half the listed number should be used for agronomic calculations.
[c] Manure plus litter at 25% moisture after 5 flocks of birds.

Disposal Systems

Today, disposal of poultry manure on cropland when conditions permit is still the best method. Several factors complicate land disposal of poultry manure. Large operations such as a 60,000-bird layer house would produce 2400 tons of manure a year. If 4 tons could be utilized per acre per year, it would still require approximately 600 acres of land, probably more land than is available on most poultry farms. Other problems include odor problems, inclement weather, equipment failures, and so on. Some producers have considered drying the manure but regardless of the method used, it involves the use of high-priced fossil fuels in the majority of situations. In Table 16.4 various systems of handling waste are categorized.

Lagoons are being successfully used in areas of the country in which weather and soil types permit. Two-stage lagoons that allow recycling of portions of the waste water are the most successful. Lagoons must be designed and managed properly to prevent odor and potential surface water contamination and, eventually, the contents must be spread on the land. Oxidation ditches have had limited success and are generally costly to install.

As the technology of waste disposal develops further, on farms, low cost poultry waste digesters should become feasible for handling cage layer waste. The methane from this process can be used as a fuel source and the digester sludge has fertilizing as well as soil-conditioning value.

Attempts have also been made to dry cage layer wastes and use them as a feed ingredient for ruminant animals. Feeding trials have shown that it can be a useful feed ingredient because microorganisms in the rumen can convert the nitrogen in uric acid to amino acids and then to protein. The costs incurred in drying it for use as a feed ingredient have prevented widespread use up to the present time.

Litter containing the manure from floor-reared chickens and turkeys currently is removed from houses about once a year and on occasion every 2 years. Poultry litter is primarily utilized by applying it to cropland and pasture. In recent years there has been growing interest in utilizing broiler litter as a feed ingredient for ruminant animals, particularly cattle. When litter is used as a feed ingredient, grain such as corn is mixed with it to provide an energy source. Molasses is then added to increase the initial palatability. The litter and grain can be ensilaged or mixed prior to feeding. Currently, litter cannot be sold as a feed ingredient. Because of variability in the product, it is difficult to meet minimum ingredient labeling requirements such as moisture level, minimum protein, and maximum fiber.

TABLE 16.4. Confinement of Laying Hens, Pullets, and Broilers

Type of confinement	Frequency of cleaning and ventilation	Manure handling and management
Caged deep dry pits	Remove manure when necessary so as to avoid crises situations; provide mechanical ventilation to keep manure dry; avoid water spillage and seepage in pit	Spread preferably when weather, soil, and cropping conditions are favorable; provide about 2 acres for each 1000 birds for land spreading[a]
Caged shallow dry pits	Scrape to spreading equipment, storage pit, or approved disposal facility as needed; provide mechanical ventilation to keep manure dry	See above
Caged deep wet pits	Remove manure when necessary so as to avoid crises situations; provide mechanical ventilation; anaerobic pits not advisable near populated areas	See above
Caged shallow wet pits	Scrape to spreading equipment, storage pit, or approved disposal facility as needed; provide mechanical ventilation; not advised near populated areas	See above
Caged aerated pits	Remove manure when solids reach 4–7% or less; advised for wet pits in populated areas	Has very little odor, therefore can be spread almost any time or place except where leaking or runoff will reach streams or bodies of water
Uncaged bedded on litter	Provide sufficient bedding to absorb moisture; remove pack as needed; maintain good ventilation	Same as for caged deep dry pits

Source: Anonymous (1974).
[a] This allows about 20 tons of manure per acre, but it may vary depending on the crop grown, soil type, and rainfall.

However, litter can be fed to ruminants by the producer. Care should be taken to ensure that no additives, such as coccidiostats or antibiotics that are under the control of the Food and Drug Administration, cause illegal residues in the carcasses of the animals fed litter. Litter has little potential for methane generation because of low yields caused by the high concentrations of lignin or cellulose from the litter source.

Hopefully, in the near future more biological methods of waste treatment can be developed. In the future, regardless of the system used and with larger units being built, waste handling has the potential to be a major problem in the path of expansion of our industry. Waste handling and pollution control will become a significant cost in the production costs of poultry.

Rodent, Pest, and Predator Control

Sooner or later in most poultry operations, rodents, pest, and predator control become important management priorities. The best management system is control rather than waiting for populations to develop to the point at which extermination is necessary. Control of rodents consists of proper design of buildings, adequate foundations in houses, and elimination of as many attractants as possible such as spilled feed, high grass or weeds, and other hiding places such as brush or lumber piles. After attractants have been eliminated, an approved rodenticide should be used carefully according to manufacturer's directions.

Pest management should also be a part of a total management plan aimed at pests such as sparrows, starlings, and pigeons, which should be controlled in areas in and around commercial poultry. Poultry buildings should be as birdproof as possible. Unwanted bird populations are a potential source of disease and parasite infestation of poultry. In addition, they consume poultry feed, can cause equipment failures, and may interfere with proper ventilation by plugging inlets with nesting materials.

Predator control consists of keeping populations under control in the area, elimination of hiding places, confinement of birds at night, use of all night lights, and trapping or poisoning when necessary.

In summary, an effective rodent control program restricts shelter, food, and water along with trapping and the use of approved rodenticides. Because of the size of modern commercial poultry houses, poisons appear to have the most potential. Approved rodenticides include anticoagulants such as warfarin and brodifacoum. Zinc phosphide is another potent poison that should be handled carefully. A relatively new rodenticide is a chemical α-chlorhydrin. Sublethal doses produce sterility in males and large doses kill either sex. α-Chlorhydrin is a restricted-use rodenticide.

Another product on the market is an adhesive glue that can be put on cardboard and placed in highly traveled rodent paths that adheres to the rodents feet with such tenacity that they become trapped in it. Regardless of whether poisons or traps are used, care should be taken to ensure that they are placed away from poultry, pets, and children. Emphasis should be on the use of a deliberate, sustained, systematic program of control measures.

Insect Control Programs

Insect control on poultry farms is a growing problem. As production units become larger, control programs become more complicated. Fly control is particularly important because of potential nuisance complaints from neighbors. Insecticides are available to be used as residual sprays, baits, larvacides, and space sprays. Generally, a combination of two or more types of insecticide is necessary to obtain effective control.

Regardless of the insecticide used, the following procedures will prove helpful:

1. Utilize insecticides as early in the season as possible to prevent population buildups.
2. Reduce fly breeding sources in and around poultry buildings. Added ventilation to lower moisture levels of droppings is advisable.
3. Use only insecticides currently approved for use in and around poultry houses.
4. Follow the label instructions for dosage rates and method of application.
5. Wear protective clothing and a respirator while spraying or preparing insecticides.
6. Dispose of empty insecticide containers properly.
7. Label and store unused insecticides safely.

In the future more emphasis will be placed on expanded control programs such as fecal larvacides as they receive approval for use. Considerable work has also been initiated with biological control measures such as parasitic wasps. The goal of insect control programs should be effective control at minimal cost with maximum safety in application and freedom from residues in poultry and poultry products.

Pesticides, Herbicides, and Sanitizers

Modern poultry farmers must rely on agricultural chemicals in their management programs. All of these compounds are harmful to humans. The only difference in their selectivity is the fact that it takes a much larger dose to kill a human than the target plant or animal. If an employee is injured his employer can be held liable. For this reason, the employer is responsible for seeing that such compounds are stored, handled, and applied in a safe manner. Nicholas Turkey Breeding Farms make the following safety recommendations.

1. Know the basic rules for safe handling of pesticides.
2. Be familiar with pesticide worker safety regulations.
3. Know where an "Emergency Care" list is posted and know which hospital handles pesticide poisoning cases.
4. Have a knowledge of the particular chemical being used when involved with disinfection or spraying of pesticides, herbicides, and disinfectants.
5. Pesticides should always be stored away from children in a locked dry cupboard or shed, in their original labeled container separately from any other compounds. Lids should be on tightly and empty containers should never be reused, but they should be held in the storage area until ready for disposal.
6. Pesticides should be prepared and applied strictly and carefully according to the label directions. It is a good practice to apply pesticides on the upwind side so they will blow away from the operator. Care should be taken to see that other individuals or livestock do not become contaminated.

Disposal of Dead Poultry

Normal mortality in poultry flocks occurs at a level of 0.5–1.0% per month in most flocks. Therefore, as commercial production units become larger, the need for daily disposal of dead poultry becomes critical. A satisfactory disposal system provides an inexpensive method to handle dead birds that prevents the spread of disease, prevents flies from breeding, and reduces odor problems while meeting state and local requirements. In recent years increasing concern over water and air quality has placed limitations on location and performance of facilities for disposing of dead animals.

Selection of a particular disposal system is usually determined by regulations in the area and will vary in different areas of the country. Poultry producers unaware of the requirements for disposing of dead poultry should contact the State Veterinarian. Basically, four methods of disposal are available when approved: burial, disposal in a solid putrescible waste cycling cell (disposal pit), incineration, and delivery to a rendering plant.

Burial In areas in which burial is allowed the requirements generally call for burial at least 3 ft deep within 24 hr of death.

Solid Putrescible Waste Cycling Cell Disposal pits, though inexpensive to operate, are difficult to construct in compliance with present requirements for excluding surface water and protecting against ground water contamination. The basic requirements from the Division of Environmental Management are as follows:

1. Locate pits at least 300 ft (100 m) from the nearest water supply well.
2. The maximum depth shall not exceed 15 ft (5 m).
3. The depth completed shall be at least 5 ft (1.6 m) above the water table.
4. The depth shall be at least 5 ft (1.6 m) above bedrock.
5. The pit must be cased from land surface to a depth of at least 5 ft (1.6 m) below the land surface.
6. Cement grout of at least 3 in. should be placed around the casing from the land surface to a depth of at least 5 ft (1.6 m).
7. The top of the casing is to be completed at least 8 in. (20 cm) above the land surface.
8. The cell must be covered with an air-tight cap.
9. The completed cell must comply with all state and local agency requirements.
10. A site inspection must be completed prior to and after inspection.

Incineration Incineration is used to a large extent for dead bird disposal. The Environmental Management Commission has specific requirements for the installation and performance of incinerators and a permit is required for each installation. Commercial incinerator installations are checked for overall effectiveness including emission control.

Recycling as a Feed Ingredient Recycling dead poultry through a rendering plant is another possibility for disposal. It has been used when the volume of dead birds is large or when other systems have become too expensive or ineffective. Recycling also allows the recovery of protein concentrates and fats. Many producers utilize deep freezers as storage for dead birds until they can be picked up by a route truck from a rendering plant.

Avoiding Nuisance Complaints   Most poultrymen can profit from improving their image as food producers.

The North Carolina Poultry Extension Service lists the following suggestions for the maintenance of grounds and buildings on poultry farms:

Keep buildings in good repair and painted, stained, or whitewashed.

Keep driveways in good condition, well drained, and holes filled.

Provide good water drainage around buildings.

Screen wild birds from inside house.

Keep areas exposed to view free from unsightly accumulations of unused equipment and junk. Store unused equipment and supplies in an out-of-sight location or storage building. Put worn-out equipment and junk in a landfill.

Landscape with grass, trees, flowering shrubs, and plants. They effectively screen buildings and unattractive areas.

Keep weeds and grass mowed around buildings and driveways. This aids air circulation and keeps the area around the house cooler and reduces erosion as well as contamination.

Adopt and practice good fly, rodent, and waste control programs.

Provide one or more garbage cans with lids for each laying house to hold dead birds until they are incinerated or buried.

Avoid littering public roads, polluting ponds or streams when hauling, applying to land, or disposing of poultry manure.

Keep odors to a minimum by good manure management. Keep manure dry.

Do not let water from fountains or buildings form puddles in which flies and mosquitoes can breed. Keep watering equipment repaired and operating properly.

Keep egg rooms clean and orderly. Wash the room daily and disinfect as needed.

Dispose of loss eggs. Keep in a covered can until discarded.

Promptly burn worn-out cases and filler flats and other trash in an approved incinerator.

Locate the incinerator away from the egg room and screen it and/or the disposal pit area by shrubbery, rail fences, etc.

Maintain the incinerator in proper repair and working order, including the afterburner. This drastically reduces odors.

Poultry producers should take pride in their poultry operations. Poultrymen whose operations are well managed and who pay attention to detail have fewer problems with their neighbors because they are good neighbors.

FINANCIAL MANAGEMENT

Production Contracts

When vertical integration became a part of the modern poultry industry, it greatly changed the business aspects of all segments of the industry. Today, the majority of broilers, turkeys, and commercial layers are produced under contract or marketing agreements. The poultry industry has evolved into a capital intensive high-volume, low-profit margin industry. Individual poultrymen limit the risk involved in production by participating in cooperation with an integrated firm in the production of poultry. Basically in contract poultry production, the producer furnishes the housing, equipment, labor, and utilities. The contractor furnishes the birds, feed, medicants, and supervision, and markets the poultry or eggs while paying the producer on some basis for his contribution.

For those individuals starting or expanding a poultry operation, the decision to construct facilities or change an independent operation to a contractual basis requires serious and thorough examination. The decision to produce poultry on contract needs to be made on the basis of economic considerations, and the interest level of the individual and his ability and dedication to make the operation a success. Any potential producer should estimate the long-term costs and returns of a contract. The interest rates should also be given close scrutiny and the depreciation schedules and investment tax credits should be considered. A check should also be made to see that all costs are included in the planning. Producers often overlook such things as wells, roads, electric service, and waste disposal costs.

Contracting Decisions

The question one must answer is whether the producer and his family have the desire, interest, and ability to make contract poultry production successful.

The North Carolina Poultry Extension Service raises the following questions for new producers.

1. Will the producer be satisfied entering into a contractual arrangement with another person or firm?
2. Will the producer accept someone else's decisions and management practices?
3. If the poultry belongs to someone else, will I manage to the best of my ability?
4. Will the producer and his family be happy with a 7-day-a-week responsibility?

5. Will the producer realize and recognize that each flock may vary in performance and therefore in returns?

Once the decision has been made to draw up a contract, both parties should study the terms carefully and take special care to see that the following points are covered:

1. *Parties and Addresses.* Is the correct identity of each party stated? Is the name and address of each party correctly stated?
2. *Execution of Contract.* Is it signed by all parties? Is it properly acknowledged and witnessed? Is the date of execution stated? Is the address of execution stated? Does each party have a copy?
3. *Contract Duration.* Is the length of the contract stated? Are the options for renewal clearly stated? Are both starting and terminating dates stated?
4. *Relationship of Parties.* Is the general relationship between the parties clearly stated? If the contractor takes over the commodity while in production, are the rights and obligations of the contractor and grower clearly stated? Does either party have the right to cancel the contract? What are the stipulations of cancellation?
5. *Facilities Used.* Is there a legal description of facilities to be used? Do both parties agree on the facilities to be used in performance of the contract? Is there a minimum size flock stated?
6. *Insurance and Taxes.* How much and what type of insurance must each party have? Who is responsible for payment of taxes on real estate, inventory, personal property, sales, and income?
7. *Arbitration of Disputes.* Are there provisions to settle disputes?

REFERENCES

Anonymous. 1974. Guidelines for Environmentally Sound Management of Animals and Animal Production Farms. NY State Dept. of Environmental Conservation. Albany, NY.

Anonymous. 1978. Turkey Breeding Manual. Nicholas Turkey Farm, Sonoma, CA.

Cooperative Extension Service. 1972. Poultry Manure—A Valuable Fertilizer, Fact Sheet 39. University of Maryland, MD.

North Carolina Agricultural Extension Service. 1973. Livestock and Poultry Waste Disposal Control, Circular 556. Raleigh, NC.

North Carolina Agricultural Extension Service, 1978. Tentative Guidelines for the Design, Installation and Operation of Animal Waste Treatment Lagoons in North Carolina, AG-18. Raleigh, NC.

North, M. O. 1984. Commercial Chicken Production Manual, 3rd Edition. AVI Publishing Company, Westport, CT.

Ryan, C. B. 1978. Texas A&M Poultry Science Laboratory Guide. Texas A&M University, College Station, TX.

Index

A

Additive gene action, 59
Albumen
    mucin fibers, 37
    proteins, 37
    structure, 37
Alectors chukar, 254
Allele, 54
Alphabatobious diaperinus, 147
Anas platyrhynchos, 227, 230
Animal liberationists, 94
Animal rights, 92
    rightists, 93
    welfarists, 93
Anser anser, 240
Anser cygnoides, 240
Argas persicus, 148
Ascaridia galli, 148
Aspergillus flavus, 145, 235
Aspergillus fumigatus, 144
Aspergillus ochraceus, 146

B

Bacterial diseases
    colibacillosis, 137
    fowl cholera, 137
    fowl typhoid, 139
    infectious coryza, 137
    omphalitis, 139
    paratyphoid, 138
    pullorum, 139
Barred Plymouth Rock, 2
Belling, 167
Behavior
    animal, 85–96
    animal organization, criteria, 92
    bonding, 90
    cannibalism, 149
    communication, 88–89
    domestic poultry, 89
    dominance system, 90
    feral chickens, 86
    growing chicks, 90

hysteria, 149
imprinting, 90
peck order, 86–87, 91
stimuli, 86
streaming, 91
Blackhead, 260
Blastoderm, 42, 65
Blood, 21
spots, 36, 41
Bobwhite quail
breeders, 253
feed consumption, 255
nutrition, 254
Body temperature, 22
Breeding systems
crossbreeding, 59
inbreeding, 58
mass selection, 60
progeny testing, 60
reciprocal recurrent selection, 58
strain crosses, 58
selection, 59
Bronchitis, 42
Broilers
brooding, 182–184
catching and hauling, 190–191
chromosomes
X, 50
Y, 50
commercial-type cross, 61–62
consumption, per capita, U.S.,
179
contracts, 192
density, housing, 184
diets and feeding, 190–191
exports, 179
feeding equipment, 186–187
hatching eggs, 180
history, 173–176
housing, 179–180
lighting, 188
litter, 182
management suggestions, 185
sex, 49
United States, 177, 178

C

Chukar partridge
breeders, 255
incubation and brooding, 255
requirements
feed, 254

space, 254
water, 254
Cimex lectularius, 147
Classification, poultry
American, 2
American Poultry Association, 3
Asiatic, 4
commercial, 5
Mediterranean, 7
utilitarian concept, 61
Claviceps purpurea, 146
Cloaca, 39
Colinus virginianus, 253
Columba livia, 261
Commercial egg production
brooding, 198
characteristics, industry, 194
crowding, 200
feeding, 200
gathering eggs, 202
manure, 202
molting, normal, 204, 205
pullets, replacement cost, 197
restricted feeding, 199
starting pullets, cages, 197, 199
water, 202
Coturnix coturnix, 249
Coturnix quail, 249
breeders, 251
brooding, 252, 253
housing and equipment, 250
incubation, 252
laboratory animals, 250
nutrition, 252
space requirements, 251
Crop, chicken, 24
Cryptoxanthine, 42
Crystal layer, 39
Cuticle, 39
Cygnus atratus, 263
Cygnus olor, 263
Cytoditis nudus, 147

D

Darkling beetles, 147
Dead poultry disposal
burial, 282
incineration, 282
problems, 281
recycling for feed, 282
solid putrescible waste cycling, 282
DNA (deoxyribonucleic acid), 51

DNA-RNA sequence, 55
Dermanyssus gallinae, 147
Disease
  definition, 126
  immunology, 130
  postmortem diagnosis, 128–129
  prevention, 127
Disease control
  vaccination, 131–132, 133
    methods, 132
    programs, 132
  vaccines, 132
Disposal systems, manure, 277
Ducks
  Aylesbury, 230
  Bantam, 232, 233
  Black East Indies, 233
  breeder, 233
  breeder facilities, 233
  brooding, 235
  Cayuga, 231
  egg handling, 234
  eggs, 232
  egg-type, 231
  English Gray Calls, 233
  Fawn, 231

E

Echidnophaga gallinaceae,
  148
Egg
  albumen, 42
  blood spot, 166
  consumer grades, 166
  consumption, 11, 12, 13
  double yolk, 41
  frozen, 169
  incubation, physiological zero,
    66
  interior quality, 167
  hatching, selection, 66
  liquid, 169
  misshapen, 41
  ovulation, 36
  size, 40
  weight classes, 166
Egg shell color, 40
Embryo
  allantois, 69
  amnion, 69
  area pellucida, 67
  blastoderm, 67

changes
  in position, 80
  in weight and form, 73
communication, 73
chorion, 69
ectoderm, 67
entoderm, 67
gastrulation, 67
growth, 69–71
head fold, 68
mesoderm, 67
mortality, 81
nutrition, 73
primitive streak, 67
research and teaching aid, 74
yolk sac, 68
Emergence, chick, 71
Endocrine system
  adrenal glands, 29, 31
  anterior pituitary, 29, 31
  function, 29, 31
  hypothalamus, 29
  islets of Langerhans, 32
  parathyroid glands, 32
  posterior pituitary, 37
  prostaglandins, 37
Energy
  digestible energy (DE), 118
  feedstuffs, 123
  gross energy (GE), 118
  metabolizable energy (ME), 118
  net energy (NE), 123
Epididymi, 43
Escherichia coli, 137, 139
Esophagus, 23
External parasites
  bed bugs, 147
  darkling beetle, 147
  fleas, 147
  lice, 147
  mites, 147
  ticks, 146, 148

F

Farmstead
  aesthetics, 101
  building orientation, 100
  planning, 99
  services and utilities, 99
  water supply, 100
Fats
  antioxidant, 114

Fats (*continued*)
  composition, 114
  crude fat, 114
  essential fatty acids, 114
  phospholipids, 114
Feathering, fast, 48
Feathers
  anatomy, 17
  function, 17
Feed additives, miscellaneous
  antibiotics, 125
  antioxidants, 125
  arsenicals, 125
  drugs, 125
  grit, 125
  pigmentors, 125
Fertilization, 51, 65
Financial management, 284
Franchise system hatchery, 82
Fungal and mold diseases
  Aspergillosis, 144
  Aspergillus favus, 145
  Candidiasis (thrush), 144
Furazolidone, 138
Further processing, of poultry carcasses,
    158, 159
Fusarium tricinctum, 145
Futures trading, 153

G

Gizzard, 24
Gametes, 50
General combining ability, 58
Genetic code, 52–53
Genetic stock, sources, 60
Genes, 49, 51, 52
  lethal, 63
  operator, 53
  regulator, 53
Genotype, 55
Geese
  African, 239
  American Buff, 239
  breeding, 242
  brooding, 243
  Canada, 241
  egg
    handling, 243
    incubation, 243
  Egyptian, 241
  Embden, 239
  Gray, 239
  nomenclature, 236
  Pilgrim, 239

Pomeranian, 240
Sebastopool, 239
Shelduck, 241
Toulouse, 239
White Chinese, 240
Germinal disc, 42
Glycogen, 113
Glycoprotein, 38
Goose
  anatomy, 238
  diseases, 245
  industry, 238, 239
  rations, 244
Grading
  poultry, 163
  shell eggs, 162–164
Growth, 111
Guinea fowl
  breeding, 258
  incubation, 259
  varieties, 258

H

Hatchery services
  debeaking, 83
  detoeing, 83
  dubbing, 83
  sexing, 83
  vaccinating, 83
Herbicides, 281
Heritability estimates, 57
Herterakios gallinarum,
    148
Heterozygous, 56
Homozygous, 56
Hormones
  adrenocorticotropin, 29
  calcitonin, 32
  estrogen, 35
  glucagon, 31
  gonadotropin, 35
  growth, 29
  insulin, 31
  luteinizing, 29, 36
  male, sex, 47
  oxytocin, 32
  melanotropin, 29
  posterior pituitary, 32
  progesterone, 36
  prolactin, 29
  testosterone, 36
  thyrotropic, 29
  thyroxin, 32
  vasotocin, 36

Houses
  floors, 103
  foundations, 102
  insulation, 103, 105
  roofs, 103
  trusses, 103
  turnkey, 102
  vapor barriers, 106
  ventilation, 106
  walls and ends, 103
Housing
  construction, 97
  design, 98
  functions, 97
Humane movement, 93
  ethics code, 95
  farmers, 94
Hybrid vigor, 59

I

Incubation
  hatching egg storage, 66
  history, 74–76
  pipping, 80
Incubator
  altitude, 78
  carbon dioxide, 78
  egg position, 79
  oxygen supply, 72
  operation, 72
  relative humidity, 77, 78–
    79
  temperature, 76–77
  troubleshooting, 81
Insect control programs, 280
Internal parasites
  capillaria worms, 149
  cecal worms, 148
  gapeworms, 149
  large roundworms, 148
  tapeworms, 149
Intestine, 25
Isthmus, 38

K

Keratin, 38
Knemidoctopes mutans, 147

L

Lentogenic form, of Newcastle disease,
  141
Light, effect on testes, 47

Liver, 25
Liquid egg processing, 159

M

Male reproduction system,
  44
  phalli, 43
  vas deferentia, 43
Mammillae, 38
Manure, 276
  handling, 278
Market news, 153
Marketing, functions of
  assembling, 151
  assuming risk, 152
  grading, 161
  inspection, 161
  merchandising, 170
  packaging, 168
  processing, 154
  storage, 160
  transporting, 152
Mating, 46
Mechanical deboning, 156
Medullary bone, 39
Meiosis, 50–51
Meleagris gallopavo, 206
Mesogenic form, of Newcastle disease,
  141
Mineral oil, 39
Minerals
  calcium, 116, 124
  copper, 124
  essential, 116
  iron, 124
  micro-, 116
  magnesium, 117
  macro-, 117
  phosphorus, 116
  potassium, 117
  salt, 124
  selenium, 124
  sodium, 117
  sulfur, 117
  supplements, 117, 124
  trace elements, function, 117
  zinc, 124
Mitosis, 50
Muscle, anatomy, 19
Mutations, 54
Mycoplasma gallisepticum, 140
Mycoplasma meleagridis, 140,
  210
Mycoplasma synoviae, 140, 210

N

National Poultry Improvement Plan, 81
Newcastle disease, 42
  lentogenic form, 141
  mesogenic form, 141
  velogenic form, 142
Nitrogen-free extract, 114
Nucleotides, 52, 54
Nuisance complaints, 283
Numida meleagris, 257
Nutrients, 112
Nutritional deficiency diseases, 132–133, 134–135

O

Omnivores, 110
Operon, 54
Oviduct
  infundibulum, 37
  isthmus, 38
  magnum, 37
Ovomucin, 38

P

Pallisade layer, 39
Pancreas, 25
Parthenogenesis, 63
Pasteurella multocida, 137
Pavo cristatus, 259
Pavo muticus, 260
Peafowl, breeding, 259
Penicillium citrinum, 146
Penicillium pupurogenum, 146
Penicillium rubrum, 146
Penis, 43
Pesticides, 281
Phasianus colchicus, 255
Pheasants
  breeder, 256
  brooding and rearing, 256
  egg production care, 256
Phenotype, 55
Pigeons
  breeder facilities, 261
  breeder pigeons, 261
  breeds, 261
  characteristics, 261
  feeding, 261
  management, 261
Pipping, 71
Porphyrin, 38

Poultry
  careers, 13
  consumption, 13, 10, 12
  distribution, 5
  enzymes, 26
  history, 2
  integration, 9
  judging, 275
  laboratory, 6
  production, 6
  respiration, 22
Poultry industry
  integration, changes in, 6
  production costs, changes in, 9
  United States, 7
  world, 6
Poultry management
  artificial insemination, 273, 274
  banding, 272
  beak trimming, 271
  brailing, 273
  brooders, 269
  brooding systems, 269
  caponization, 274
  clipping, wing, 273
  desnooding, 272
  dubbing, 272
  feeders
    space requirements, 270
    types, 266
  leg banding, 272
  litter, 268
  pick guards, 273
  pinioning, 273
  sexing, 274
  toe clipping, 272
  water consumption, by chickens, turkeys, 271
  waterers
    space requirements, 270
    types, 267
Processing functions, poultry, 154–137
Product development, 172
Protein
  amino acids, 115
  animal and vegetable supplements, 123
  biological value, 115
  calculation, 115
  classification, 115
  hydrolysis, 115
  nitrogen, 115
Proteus, 139

Protozoan diseases
blackhead (histomoniasis),
136
coccidiosis, 133
hexamitiasis, 136
trichomoniasis, 136
Proventriculus, 24
Pseudomonas, 139

R

Recessive traits, 65
Reproduction, 65
Reproduction physiology, 33
female, 33
light, effects of, 33
organs, 33
Reproduction diseases, oviduct
impacted, 150
prolapse, 150

S

Salmonella, 139
Salmonella gallinarum, 139
Salmonella pullorum, 138
Sanitation, hatchery, 82
Sanitizers, 281
Selection, 57
Semen
collection, 46
storage, 46
Sequence, egg laying, 40
Sexing, 47
Sex-linked inheritance, 57
Shell egg processing
flash candling, 159
sizing, 159
storage, 159
washing, 159
Shell membranes, 38, 39
Shell thickness, 43
Showing, of poultry, 275
Skeleton
calcium requirements,
21
development, 19
Skin
anatomy, 18
color, 18
pigments, 19
Sperm, turkey, 45
Standard of perfection, 60
Stigma (suture line), 36

Strain crossing, 61
general combining ability, 58
Structure, egg shell, 43
Sulfa drugs, 138
Suture line, *see* Stigma
Swan
Black Australian, 263
breeding, 264
breeds, 263
characteristics, 263
incubation and brooding, 264
Trumpeter, 265
White Mute, 263
Syngamus trachea, 149

T

Thermometer
dry-bulb, 77
wet-bulb, 77
Treading, 46
Trichophyton gallinae, 145
Trombicula alfreduggesi, 147
Turkey
Austrian White, 207
Bourbon Red, 207
brooder guards, 217
confinement rearing, 221–222
consumption, 225
contract producers, 223
equipment arrangement, 218
housing, cost of
building, 224
equipment, 224
range, 224
mechanical feeders, 220
Narragansett, 207
origin, 206
production, 225
range rearing, 220–221
trough feeders, 220
tray feeders, 219
waterers, 220

U

USDA grades and standards, 162

V

Vagina, 38
Velogenic form, of Newcastle disease,
141
Ventilation
forced-air, 107

Ventilation (*continued*)
  negative-pressure, 108
  positive-pressure, 109
Viral diseases
  avian influenza, 143
  fowl pox, 142
  infectious bronchitis, 142
  infectious bursal disease, 142
  infectious laryngotracheitis,
    142
  lymphoid leukosis, 143
  Marek's disease, 143
  Newcastle, 141
Vitamins
  poultry requirements, 119–
    120
  supplements
    A, 125
    B_{12}, 125
    choline,
    D, 125
    niacin, 125
    pantothenic acid, 125
    riboflavin, 125

W

Waste disposal, 276
Water
  bacteria, 113
  consumption, 112
  nitrates, 113
  nitrites, 113
  quality, 112
  salt, 113
  sulfates, 113
Waterfowl
  dressing percentages, 245
  handling, 247
  pinioning, 247
  processing, 245
  sexing, 246, 247
White Leghorn, 61

Y

Yolk
  carotenoid pigments, 42
  concentric layer, 42
  vitamin A, 42